To
Nat Earle

With best regards

THE PROSPECT FOR GOLD

THE VIEW TO THE YEAR 2000

THE PROSPECT FOR
GOLD

THE VIEW TO THE YEAR 2000

TIMOTHY GREEN

WALKER AND COMPANY **NEW YORK, NEW YORK**

First published in the United States of America in 1987
by Walker Publishing Company, Inc.

Library of Congress Cataloging-in-Publication Data
Green, Timothy, 1936-
The prospect for gold.

Bibliography: p.
1. Gold. I. Title.
HG289.G7395 1987   338.2'741   87-16046
ISBN 0-8027-1002-6

Produced by
Rosendale Press, London, England

Jacket & design by Pep Rieff

Typesetting by TypeArt, London, England

Printed in the United Kingdom by Biddles Ltd, Guildford, Surrey
10 9 8 7 6 5 4 3 2 1

To the memory of my mother
Phyllis May Green
1907 – 1987

CONTENTS

Preface

This book looks at the gold business in the late 1980s and asks where it is going for the rest of this century. In the twenty years that I have been writing about gold the business has already been transformed from the days of $35 an ounce, strictly maintained by the central banks' gold pool, into a free-wheeling, round-the-clock trading game. I have tracked many of those changes in earlier books, starting with *The World of Gold* through to *The New World of Gold*. *The Prospect for Gold*, by contrast, looks forward to the challenges ahead, particularly because the present boom in gold mining has opened up a real question of the price at which all this new supply can be absorbed.

I approach gold as a journalist, not an economist, and I am not attempting to expound theories on the role it should, or should not, play, as money or commodity. However, I do have pragmatic views on how the industry should confront this new era, and that is one reason for writing this book.

Many people in many countries, who have become good friends over the years, have helped my research. I have a special debt to David Lloyd-Jacob, who initiated gold market research at Consolidated Gold Fields in London in the late 1960s and offered me a territory 'east of Venice, up to and including Hong Kong'. Since then Peter Fells, Christopher Glynn, David Potts, Louise du Boulay and George Milling-Stanley, the successive editors of Consolidated Gold Fields' annual market survey, have all enabled me to keep an eye on gold. I am grateful, too, to Willem Beunderman, Andrew Golding and Jessica Jacks for much advice. Iris Flaherty, the gold survey's secretary, has worked unstintingly to keep our gold research on the road.

Also in London, I must pay tribute to Julian Baring, Robert Weinberg and their colleagues at James Capel, whose weekly *Mining Review* has focussed my attention on many new gold prospects.

Around the world, my travels for this book have been aided by a host of people. In Australia, I am grateful to Don Morley of Western Mining Corporation, Don Mackay-Coghill and his new team at Gold Corp Australia, Doug Daws who was a wonderful guide around Kalgoorlie, David Tyrwhitt at Newmont Mining, Robin Jenkins, the diligent librarian at Renison Goldfields, Ross Louthean, editor of *Gold Gazette*, and Trevor Sykes, editor of *Australian Business*, whose reporting of Australia's gold rush is outstanding. In Canada, I am grateful to Robert Calman and Paddy Broughton at Echo Bay Mines, Peter Allen at Lac Minerals, Paul Warrington at Noranda, Donn Morgan at Placer Development, and Hank Reimer and Ted Reeve at Loewen, Ondaatje, McCutcheon, together with all the miners and geologists at the Hemlo and Lupin gold fields who guided me around. In the United States, I had much help from David Fagin at Homestake Mining, Ron Zerga and his team at Newmont Gold in Nevada, Philip Lindstrom at Pegasus Gold, Jeffrey Nichols of American Precious Metal Advisors, and Amy Gassman at Goldman Sachs. In Tokyo, I must pay special tribute to the long-standing help of Juni-chiro Tanaka, Takashi Tanaka and Tadahiko Fukami at Tanaka K.K., and also to the advice of Yoshio Sekine at Tokuriki, Seuchi Komiya at Ishifuku, Akio Imamura at Sumitomo Corporation, and Itsuo 'Jeff' Toshima at the World Gold Council. In Hong Kong, I have long valued the advice of Kenneth Yeung of King Fook Finance and Robert Sitt of Samuel Montagu. In India, my good friend Madhusudan Daga continues to be my guide, together with Shantilal Sonawala, president of the Bombay Bullion Association.

The advice and friendship of many other bankers and bullion dealers in such diverse places as London, Geneva, Zurich, Basle, Chiasso, Frankfurt, Luxembourg, New York, Toronto, Jeddah, Riyadh, Dubai, Singapore and Jakarta is also appreciated, but they prefer to remain discreetly anonymous.

Finally, my special thanks go to my wife Maureen, who has edited this book, to Pep Rieff who has designed it, to Judy Walker who copy-edited and to Clare Martin and Georgie Robins at Housemartin who typed it.

T.S.G.
Dulwich, July 1987

Golden Rules

WEIGHTS

Gold is measured in either ounces troy or metric units. I have normally used the metric system, except in mining in North America and Australia, where producers usually quote ounces and I have maintained that custom, with the metric tonne equivalent in brackets. A few basic conversions may be helpful.

1 oz troy	=	31.10 grams
32.15 oz troy	=	1 kilo
100 oz troy (Comex contract)	=	3.11 kilos
32,150 oz troy	=	1 tonne

Special bars

10 tolas (Indian sub-continent)	=	3.75 oz or 116.6 grams
1 tael (Hong Kong)	=	1.2 oz or 37.3 grams
1 baht (Thailand)	=	0.47 oz or 14.6 grams

PURITY

The purity of gold is measured either by the carat (karat in USA) or by its 'fineness' in parts per 1,000. The gold market does not recognise a higher purity than 9999 (four nines), but gold wire for electronics may be refined to five nines.

Fineness		*Carat*
9999	=	24
916.66	=	22 (some coins and 'investment' jewellery)
750	=	18 (much European jewellery)
583.3	=	14 (most North American jewellery)
375	=	9 (most UK jewellery)
333.3	=	8 (normally lowest acceptable in jewellery)

Introduction

THE PROSPECT FOR GOLD

The first question on everyone's lips in the gold business nowadays is 'What will the price do?' Bankers and bullion dealers, coin dealers, jewellery manufacturers, investors and even gold smugglers inevitably have that concern uppermost in their minds. Yet it is a relatively new question. The historic reputation of gold was founded on the very stability of its price. It was the bench mark *par excellence* against which other prices were compared. The sterling price of £3.17s.10½d. per standard troy ounce set by Sir Isaac Newton in 1717 lasted for over 200 years, with only a few minor fluctuations (like the days after Napoleon's escape from Elba). President Franklin Roosevelt also decreed $35 an ounce in 1934 and that, too, lasted a generation. So gold traditionally was, in Roy Jastram's phrase, 'The Golden Constant'.[1]

Today, it is a different story. The last few years have seen gold soaring to $850 per troy ounce and tumbling back under $300; often the price fluctuates $10, $20, even $30, in a few hours, and exceptionally $100 in a single day. The game has changed. Many people now buy gold, not as a safe haven whose value is rock solid, but hoping to make money.

Thus, setting out on an odyssey around the gold business in the late 1980s and trying to foresee where gold may be headed for the rest of this century and beyond, we must accept that the whole environment has altered. A tidal wave of investors' money switching momentarily out of bonds, equities or currencies can engulf the relatively small gold market (whose new supplies amount only to about $20 billion annually if gold is at $400) in an afternoon. Given such a switch from time to time, one cannot rule out prices of $1,000-$2,000 or even beyond before the year 2000.

Equally, one should not ignore that this decade has ushered in an almost unprecedented new wave of exploration for gold that is lifting production to levels undreamed of a few years ago. Gold mining has become the last really profitable mining game in town. If investors are not inclined to favour gold, we could be heading for oversupply in the 1990s, unless radical new initiatives can stimulate the regular 'bread-and-butter' demand for jewellery, industry and coin. A scenario of no increase in the gold price in real terms to the end of this century

1 Roy Jastram, *The Golden Constant*, John Wiley, New York, 1978

can be painted as easily as one projecting $1,000 or more.

Many other forces buffet the gold price up and down. What will happen in South Africa? Revolution might seriously disrupt the source of half of non-Communist gold for several years, sending the price shooting up. In the same breath do not forget China. She is slowly emerging as a significant gold producer. As her own population of a billion acquire more wealth, will they provide a much-needed new demand? The Russians, among the most wily players of the gold market, also love to hint that they too have lots of new deposits, and under the pragmatic Mr Gorbachov might become much larger sellers to pay for urgent imports. The Japanese, too, played a decisive hand in 1986, picking up 650 tonnes of gold, but is their appetite limited? Judging the gold price is like riding a roller-coaster; it dips when you least expect it, and soars apparently against all reason.

Technical chartists now explain it all with delightful double bottoms and heads and shoulders. The most eloquent technical chartist, Brian Marber, loves to talk of little clouds of gnats dancing in the evening sunlight in the forest which move all of a sudden *en masse* from one spot to another, seemingly without a breath of wind: just so, speculators suddenly dart into or out of the market. Alternatively, Bob Prechter in the *Elliott Wave Theorist* newsletter expounds on the natural rhythms of investment psychology and their impact on markets. Such 'waves' were first deduced in the 1930s by a Los Angeles accountant cataloguing historical market movements.

Chartists and their ilk were greeted initially with some scepticism when they first tackled gold, but now market makers ignore them at their peril. 'It's often a very, very technical market,' admitted a major operator for Middle East clients. 'On Comex the big US funds and computer operators all come in and out at certain points and we have to advise our Arab clients, who used to buy or sell on sentiment, to operate on a technical basis.' Computer traders, with programmes automatically giving buy or sell signals at key chart points, can dictate the direction of the market in the short term. Woe betide any grand speculator out on a limb with a massive short position in Comex futures contracts if the price moves against him or the writer of 'naked' options to provide gold when he does not own one ounce.

Financial disaster has descended on the unwary. Abdul Wahab Galadari, the Dubai merchant and banker, lost close to $100 million through a staggering short position of 1.5 million ounces on Comex in August 1982 as the price took off from $330 and finished at $507. The last $100 on that rise was triggered just covering Galadari's position. An Egyptian speculative group, the Rayan Company for Investment, was reported to have taken a similar beating in September 1986, when they were one million ounces short and gold slid up over $400 and kept going. 'We had an order to sell 300,000 ounces for one client at $400,' a London dealer told me, 'and we thought that would cool the market, but it was just snapped up without a murmur in the Rayan covering.'

Such volatility may be adored by the major New York commission houses like Merrill Lynch, E.F. Hulton or Prudential Bache, which thrive on an active market, but it cripples the traditional jewellery manufacturer. He needs a steady price. Industrial users, too, may look for more stable alternatives. While gold has not suffered the fate of silver, whose fabrication base, especially in sterling silverware, was virtually destroyed by the $50 price explosion triggered by Banker Hunt in 1980, there has been a decline in industrial use as high technology gets by with micro-dabs of gold on contact points and semi-conductors.

Moreover, many regional gold markets of the Middle East and South-East Asia have shown remarkable sensitivity to sudden price movements. They go into reverse virtually overnight if the price soars and start dishoarding gold back to London or Zurich, often in significant amounts. The price surge in the autumn of 1986, for example, prompted dishoarding of over 100 tonnes of gold inside an eight-week period, and more was triggered by the high prices of early 1987. These sudden reverses from markets that may previously have been absorbing 50-100 tonnes a month do much to cool speculative flurries.

This may be the era of technical charts, but the chart I am presenting for the first time in this book (see Chapter 13) on the ebb and flow of physical gold to key regional markets in the Middle East and Far East demonstrates how the fundamentals of supply and demand still impose themselves on price trends over a period of time. My chart indicates clearly that in early 1983, for instance, with gold around $500, the market was in highly speculative territory with virtually no physical demand and heavy dishoarding.

At that moment technical chartists and many bullion dealers cheerfully predicted $600 within weeks. In fact, with no real gold demand the price duly collapsed back towards $400. Equally, in the first two months of 1985, with gold at under $300, the physical offtake was far exceeding normal supplies coming to the market and it was only a matter of time before the price took off upwards. Then in the autumn of 1986 the price spike to $440 again eroded physical demand, triggered dishoarding, and the price eased back under $400.

Do not be mistaken. I am not saying that the fundamentals of supply and demand are sole arbiters of the price. Far from it. Investment sentiment can be much more important. Fear of inflation or a weak dollar can push the gold price up in a way physical demand rarely can (though the exceptional Japanese offtake in 1986 certainly underpinned and improved a weak market). Monitoring the changing physical flows, however, can signal when gold is in 'safe' territory, with a high physical offtake providing a floor, or when it is in speculative waters with no 'bread-and-butter' demand. Thus, 'buy' at $290 in February 1985, 'sell' at $440 in October 1986. Such signals may not help the short-term trader, who is in this morning for 100,000 ounces and out this afternoon on a $5 rise. They can give confidence, however, to investors with a longer view of gold, and also to gold miners looking to see a realistic floor for the price, so that they can determine if a new gold mine prospect is worth developing. The latter decision has become quite easy in recent years, with gold averaging between $300 and $450, while miners find plenty of new prospects for development at $200 or below.

The renaissance in gold mining is one of the major themes of this book. We are witnessing the greatest explosion in gold production since the Californian and Australian gold rushes around 1850 and the South African discoveries of the 1880s opened up an entirely new dimension to annual gold supplies. In the second half of the nineteenth century, three times as much gold was mined as in the first half. Today's growth may not be quite so rapid, but between 1980 and 1987 non-Communist mine suppliers rose by fifty per cent. Further expansion is forecast until well into the 1990s. Already Canadian output has doubled, United States has quadrupled and Australian is up six-fold. Over eighty per cent of the exploration budgets of mining companies worldwide are now devoted to the hunt for gold. They are finding it: in Egypt, Sudan and Saudi Arabia, in Ivory Coast and Guinea, in Papua New Guinea, the Solomon Islands and Indonesia (where a world-class mine is due in 1989). In Japan, Sumitomo Metal Mining has unveiled a unique silica mine with gold grading to 150 grams per tonne (versus average 6 grams a tonne in South Africa). Throughout Latin America, alluvial gold is being scoured out by hundreds of thousands of diggers from such El Dorados as Serra Pelada and Cumaru in Brazil, or Kilometro 88 in Venezuela.

The full potential of this expansion was brought home to me at the workshop on the long-term outlook for gold organised by the Centre for Resource Studies of Queen's University in Kingston, Canada, in the autumn of 1986. In an address to mining company chief executives, bankers and bullion dealers, Jim Fisher of Brook Hunt & Associates, the London mining industry consultants, underlined, 'The sharp increase in real gold prices has led to an order-of-magnitude re-evaluation of the place of gold in the portfolio of mining companies' operations, to the point where gold is now the principal target of almost all metal exploration programmes. Given the increasing rate of exploration expenditure...it is difficult to make meaningful estimates of production from deposits which may be found in the future.' He drove the message home by adding that, as recently as 1980, no forecaster had predicted anything like today's level of production.

This raises the crucial question. Who will buy all this new gold? Or, more important, at what price can it be absorbed? Is there a real danger of oversupply, as in silver, or even – if one dare mention it – tin?

In the nineteenth century, the flood of metal from the gold rushes financed a great expansion in world trade and ushered in the true, short-lived, era of the international gold standard. Fears that the gold rushes would depress the price then were unfounded. This time the environment is different. Gold's fortunes today are based much more on its role as a commodity, rather than as a monetary metal. In the nineteenth century, gold was money. Now, for all the spate of new coins for investors, its cornerstone is in jewellery and industry. Within the last twenty years the balance has shifted dramatically. In the immediate post-war years, gold was still taken up largely by central banks. Between 1948 and 1965, forty-two per cent of all new gold was bought by monetary authorities.

Since then, what a different story! Nearly sixty per cent of new gold coming on the market has gone into jewellery, another fourteen per cent to industry, and a similar share to private bar hoarding, with bullion coins taking up most of the balance. Central banks were actually net sellers of gold between 1966 and 1986. Only in the 1980s have central banks emerged marginally as net buyers again. First, OPEC countries like Iran, Libya, the United Arab Emirates, Qatar and Indonesia put some of their oil wealth into gold in the early 1980s. But the collapse of oil from $35 a barrel soon ended that. Then countries like Brazil, Colombia and the Philippines started buying up blossoming local production with rapidly depreciating paper currencies, and either stockpiled the gold or sold it for hard currency. This development prompted a leading London gold trader to observe, 'We are getting onto a gold mining standard, in which countries with weak currencies produce more and more gold and sell it to countries with strong currencies like Japan.' This thesis, of course, fits not only the non-Communist world, but the Soviet Union and China too. Both trade gold for hard currencies. The Russians incidentally often sell gold for deutschmarks in which two-thirds of their imports are billed. In that sense, gold sales still help to underpin world trade, but essentially as a commodity. The Russians view gold as a foreign exchange earner, just as they do oil, arms, diamonds and furs.

The test for gold for the rest of this century, however, is going to be how well sustained the jewellery, industry and coin demand will be. This 'bread-and-butter' demand provides the real floor for the gold price. As production soars, so the 'bread-and-butter' offtake must grow to provide the essential cushion to hold the price in years when investors and speculators no longer perceive the metal as a hedge for inflation or other anxieties, and desert for other more volatile and lucrative markets. Speculators are not going to carry gold year in, year out.

The burden for the rest of this century is clearly going to fall increasingly on countries in the Pacific basin - like Japan, Taiwan, Hong Kong, Singapore, Malaysia and perhaps India, that traditional sponge for precious metals. The Middle East, whose oil wealth was such a driving force in the gold market from

the mid-1970s, until the early 1980s, is a spent force. Saudi Arabia has actually become a net exporter of physical gold back to the markets of London and Zurich
as declining oil revenues have dried up internal liquidity. Visiting Saudi Arabia
early in 1987, I found that any surge in the gold price was looked on with relief as
an opportunity to get out of gold at some small profit, or, more often, lower loss.
Over $1 billion in scrap gold left Saudi in 1986. A Jeddah gold dealer told me, 'If
the price went to $500, you'd see many people here sell up everything and close
their gold shops.' At least one OPEC government investment fund considered
selling some of its strategic stock of gold in order to realise assets in its portfolio
to compensate for less oil income.

Looking into the crystal ball for the future of the gold price for the rest of
this century, we must not forget how different the economic environment is from
the days of the late 1970s, which led to the heady $850 price in January 1980.

As *The Economist* observed, 'The darlings of the inflationary OPEC decade,
hard assets like commodities and gold and some real estate, have gone out of
favour.'[2] Will they return? Make no mistake about it, there is plenty of money
that could swing back into gold. OPEC members may be less awash with cash,
but in America and Japan individuals have literally billions in liquid assets in
money market deposits and banks, to say nothing of the funds in stock and bond
markets. If gold does come back in fashion, as it showed real signs of doing in
1987, there is no lack of money to drive it to new heights. A price of $850 is not
difficult to surpass in the right mood. No doubt that will happen long before the
year 2000. The question is how soon, and what will be the catalyst? And, between
times, how vulnerable is the price on the downside? Can miners go on finding
more and more at $200 an ounce, when the price is anything from $300 to $500?
Will the day of reckoning come?

2 *The Economist*, 24 January 1987

Part One

MINING:
A NEW ERA

Chapter 1

THE ONLY GAME IN TOWN

The 1970s was the decade of the gold speculator; the 1980s, unquestionably, has been the decade of the gold miner. The driving force of speculation in an era of high oil prices and inflation pushed the gold price to unprecedented heights in 1980, and so caught the attention of the gold miners, who had neglected the metal for a generation. Suddenly, as base metal (and oil) prices weakened, gold stood out as a profitable beacon beckoning geologists, mining engineers and, of course, bankers and investors eager to finance them. 'If you go to the bank with a gold project, you are welcomed with open arms,' said a New York mining analyst as we flew to look over a new open-pit mine in the American West. 'But if you go with a base metal project, they'll think you're out of your skull.' Chester Millar, chairman of Glamis Gold, a small Vancouver-based company, was equally blunt. 'It's the only way you can make a dollar these days,' he said. 'We'll chase gold willy-nilly. What else can we do? There's no other game in town.'

No, indeed. With gold averaging $367 an ounce in 1986 and forging on comfortably over $400 the following year, miners throughout the Western world were not only operating existing mines at costs usually under $250 an ounce, they were finding plenty of new prospects that could be brought on stream with cash operating costs under $200 and capital costs that could be paid off within a year or two. Placer Pacific's new Kidston mine, the largest to date in Australia, which opened in 1985, actually paid for itself in a little over twelve months, while Greenwich Resources has brought in a small new mine in Sudan with the astonishingly low cost of $51 an ounce. The three new mines at Hemlo, the most important gold field found in Canada this century, all have operating costs below $200. This wide margin between costs and the gold price provides the industry with remarkable insulation. Even if the price fell back under $300, the majority of mines would remain profitable and little output would be lost, at least for the first year or two.

'The profitability of gold mining over the past five years has been a major factor in persuading mining companies to increase their exposure to the metal,' observed Jim Fisher of Brook Hunt & Associates, the London mining consultants. 'For each year since 1979, over ninety per cent of gold production in the

Western world has shown an operating profit. No other metal has come anywhere near this record in recent years.'[1] While prices of such base metals as copper and zinc are down in real terms by two-thirds since 1970, the gold price, in real terms, has more than tripled.

The consequent shift in the energies of mining companies is understandable, but quite remarkable in scale. Prior to 1968, while the gold price was fixed at $35, exploration for new gold deposits outside South Africa was negligible. Even in 1975, after the price had first risen close to $200 an ounce, gold commanded barely ten per cent of exploration budgets. Today it is often eighty per cent or more. In Australia alone, exploration expenditure soared from A$30 million in 1980 to over A$200 million in 1986. Overall, close to US$1 billion is now being invested annually in the search for gold, triple the amount in 1980.

Already the world map of gold production has been radically re-drawn. South Africa, once the source of seventy-five per cent of gold outside the Communist bloc, accounted for just under fifty per cent by 1987, not because her own output was down much, but because countries like Australia, Canada and the United States were up. Those three countries contributed less than 100 tonnes of gold between them in 1980; by 1987, each individually will easily top the 100-tonne mark. Other such diverse places as Brazil, Chile, Colombia, Indonesia, Ivory Coast, Japan, Papua New Guinea, Sudan and Venezuela show good increases (or have started from scratch). And watch out shortly for New Zealand, Saudi Arabia, the Solomon Islands and Vanuatu joining the gold club. In the Western world, non-South African output has shot up from a mere 280 tonnes in 1980 to about 700 tonnes for 1987.

This world-wide revival embraces not just traditional mining houses with gold experience, like Anglo-American, Gold Fields, Homestake, Lac Minerals, Newmont, Noranda or Western Mining Corporation, but groups such as Amax, Kennecott, Placer Development and St Joe Minerals, whose principal experience was in base metals. Placer, the Vancouver-based company, only really decided to go for gold in the late 1970s, but is now firmly established in Australia and Papua New Guinea through its Placer Pacific spin-off, while in Canada it has linked up with Dome Mines and Campbell Red Lake. At Amax the decision is more recent, but chairman Allen Born has said he wants his multi-national to be a major gold producer inside a decade. (Amax has a nice new little mine called Sleeper in Nevada.) Several big houses have already spun off their gold divisions as separate companies - Newmont Gold, Freeport Gold, St Joe Gold, Hemgold - to make them more attractive to would-be investors as a pure gold share.

These established mining houses have been joined by other new faces who had previously made their names in oil, like Canada's Peter Munk with American Barrick, Battle Mountain Gold which grew from Pennzoil, and BP Minerals, a

---

1 J.F.C. Fisher, 'Mine Gold Supply to the Year 2000', paper presented to the Centre for Resource Studies, Queen's University, Kingston, Canada, September 1986

child of British Petroleum. Then there is Echo Bay Mines, whose previous experience was a small silver mine in the Canadian Arctic, which originated under the wing of the IU International investment group. Alongside the big names, a legion of small mining companies in Australia, Canada and the United States have also successfully started low-cost open-pit operations.

The players, old and new, have come up with original financing ideas. Banks and bullion dealers loan gold to miners against their future production, providing them with instant cash flow by selling the gold to get the mine going. Gold bonds and warrants give punters the opportunity to take a slice of a mine's future production, which becomes the more attractive as the price rises. Many small companies have tapped into venture capital on the Vancouver Stock Exchange, which has proved a world-wide magnet for gold share speculators, who enjoy a heady atmosphere that reminds many of Las Vegas gambling tables. More serious institutional investors have also been attracted by the larger market for gold shares created by such spin-offs as Newmont Gold or Placer Pacific from more diverse parent companies, and by houses like American Barrick and Echo Bay building up a stable of promising properties. 'We have moved from small issues of junior stocks to highly capitalised tradable securities,' Steve Dattels of American Barrick pointed out.

Yet the continued contribution of small companies, often tackling marginal projects where big mining houses (with big overheads) fear to tread, should not be under-estimated. Their diligence has often pointed the way to major finds, like the Hemlo gold field in Canada which was initially charted by small entrepreneurs. The chairman of one such company, which operates a couple of open pits in California, put it this way, 'We are lean and mean, and to make a buck we live in the garbage can of big companies.'

New Technology Aids the Open Pit

The Who's Who of gold mining has thus been completely re-written and expanded in recent years. So, too, has the gold miners' manual. Price may have been the original catalyst, but as miners focussed on gold (often for the first time in their lives) they found a range of new technology, from computer control of ore grades to radically improved recovery techniques, which enabled them to work low-grade open-pit deposits previously discarded as uneconomic. Indeed, the rapidity of the rise in Western gold mine production, from 950 tonnes in 1980 to around 1380 tonnes in 1987, is chiefly due to the explosion of open-pit operations, particularly in Australia and the United States. Open pits telescope the time and financial scale. A deep South African mine will cost anything from $500 million to $1 billion to bring into production, and seven years may elapse before an ounce of gold emerges. By contrast, in Nevada in the United States and around Kalgoorlie in Western Australia, open pits are working within a few months for the expenditure, at most, of a few million dollars. The fastest start-up I heard of, near Kalgoorlie, was twenty-seven weeks from the decision to go ahead till production began.

The viability of working many low-grade deposits has been aided by new assay techniques, enabling thousands of assays to be taken from an open-pit floor and their results translated into a computer model of the mine. 'Computer technology actually makes mines people were afraid to tackle before,' a geologist told me at Carlin in Nevada, where Newmont Gold is tapping no less than eleven orebodies along a range of hills. 'On a computer, you can now do a mine plan every month; before it took a year.' That plan targets gold-rich ore, as distinct from waste, throughout an open pit. Coded coloured flags or white-washed markings can then indicate on the pit floor where blasting or bulldozing should be done, so that good ore is separated from waste immediately. This control is critical to the profitability of the mine when cut-off grades are often less than 1 gram per tonne.

Heap-leaching Comes of Age

Two other principal technical innovations have stimulated gold output - heap-leaching and carbon-in-pulp recovery. Heap-leaching of gold was first pioneered in the United States in 1973 at Placer Development's Cortez open pit in Nevada and then proved on a larger scale at Pegasus Gold's Zortman Landusky mines in Montana. It has the prime advantage of enabling gold from low-grade deposits of 3 grams per tonne or less to be extracted economically. In conventional mining, the ore has to be crushed to a powder, often at considerable expense, and then passed through cyanide tanks to take out the gold. Heap-leaching eliminates much of this costly capital investment in milling plants. Instead, at an open pit the ore is blasted and bulldozed, put on huge trucks which heap it on open-air leach 'pads' with a base of asphalt or impervious plastic sheeting. A sprinkler system is then laid along the top of the ore and a solution of dilute cyanide is sprayed on the pile for a couple of days to saturate it. The cyanide percolates down through the heap of ore for several weeks, leaching out the gold. This solution, now enriched with gold, drains off the bottom of the pad into what is aptly called 'the pregnant pond'. 'Heap-leaching is the exact reverse of marinading beef,' a mining engineer told me in Nevada. 'And it's cheap. We got $900,000 worth of gold out of one leach pad for only $23,000 worth of cyanide.' In the United States, where heap-leaching is most widely used, over half of all output now comes from this process.

After leaching, the second new technique, carbon absorption, comes into play. The 'pregnant' solution is pumped to the recovery plant, where it passes through a series of tanks containing carbon granules. The granules are best made from coconut shells, which turn into a hard carbon durable enough for repeated use. Gold has a natural affinity for carbon, and in as little as six minutes ninety-seven per cent of the gold transfers its allegiance from the cyanide solution to the coconut-shell carbon as it passes through the absorption tanks. The carbon, in a sense, is acting like a sponge soaking up gold. The cyanide solution, no longer pregnant, can be re-used to leach more heaps. The carbon, now pregnant in its turn, goes to pressurised tanks. At great heat, the carbon expels the gold again into a fresh cocktail of sodium cyanide and caustic soda. Finally, this pregnant solution is passed through electrolytic, or 'electro-winning', cells, whose cathodes are covered with steel wool. The gold (with any associated silver) is deposited on the steel wool from which it can easily be recovered by fire refining. The carbon, meanwhile, is re-treated so it can be used again.

Taken together, heap-leaching and carbon-in-pulp recovery have revolutionised gold mining. Cut-off grades as low as 0.3 grams per tonne have been achieved on some mines, a tremendous scaling down when one remembers that South African mines average about 6 grams per tonne and the new underground mines at Hemlo in Canada are around 10 grams. Leaching, however, is not a cure-all; it has limitations. It works best in warm, dry climates (like Nevada) because heavy rainfall would dilute the cyanide and frost would freeze the ore piles. So heap-leaching is not yet applicable in wet, tropical climates like Indonesia or Papua New Guinea, or cold ones like Canada or the Soviet Union. It can also be applied only to porous ores, and even then only sixty-five to seventy-five per cent of the gold is recovered. The point, though, is that *none* of the gold could be recovered economically by conventional techniques. In practice, mines often use a combination of conventional milling and heap-leaching; higher-grade ore goes through the normal mill for the best recovery rate, while the lower grades are diverted to the leach pads. Computer control of ore grades makes that kind of selection practical from day to day.

Carbon-in-pulp recovery, by contrast, has much broader applications, not just for new open pits but for traditional underground mines, where its introduction often improves recovery grades, cuts costs and keeps old mines going. The process is widely used in North America, Australia, Brazil and other developing areas. In Australia it has proved particularly useful in enabling treatment of oxidised ores with a high clay content, which could previously be treated only in high-cost plants. Carbon absorption also works well with poor-quality and highly saline water, which could not normally be used in the customary treatment processes.[2]

The great leap forward in production owes much to these processes, and their potential is still not fully exploited. Can heap-leaching, for example, be

2 Paper by Don Morley, Director of Finance, Western Mining Corporation, delivered Johannesburg, September 1986

done economically under cover to eliminate the climatic problems? And can other types of leaching be developed? 'I believe there's a future for *in situ* leaching,' Peter Steen, chairman of Royex, the Canadian mining group, told me. 'Using oil-well technology, you pump down a cyanide solution into refractory rock, collect the [pregnant] solution in underground working and bring it up.' This would eliminate all blasting and hauling of ore.

Another long-range idea is bacterial leaching, which might be applied to sulphide ores from which the metallurgy of unlocking the gold is very complex. 'Bacteria will eat sulphide,' said Chester Millar of Glamis Gold, 'rendering rocks porous and amenable to heap-leaching.'

Seeking the Origins of Gold

Such initiatives may be several years ahead, but they demonstrate how the gold mining industry is seeking the benefits of new technology. Satellite pictures, for instance, may prove invaluable in pinpointing new orebodies. Old-time prospectors with picks and shovels have given way to university-trained geologists. 'Today there is more willingness to make sense of a theory, and then apply theoretical lessons in the search for the same kind of rock types,' said Peter Allen, president of Lac Minerals, which maintains one of Canada's best exploration divisions. 'We say, maybe two sets of rock are related, let's try.'

A great debate is going on as to the true origins of the Archean quartz vein gold deposits in belts of greenstone rock which house many of the richest fields in Brazil, Canada, the Soviet Union, South Africa and Western Australia. Half the world's gold output comes from such deposits, which were formed by a hot gold-carrying fluid pushing up from the depth of the earth into a network of fissures in the greenstones. But where did that fluid come from and what exactly was it? The accepted idea is that the gold veins came from metamorphic fluids, generated by the dehydration of another form of rocks. Three Canadian geologists, led by Dr Ed Spooner at the University of Toronto, are convinced, however, that the gold veins came from magmatic fluids welling up from molten granite-like masses in the earth's core.[3] If they are correct, then the search for new orebodies might have to be redirected. Dr Spooner believes that, if the origins of the gold are indeed magmatic, then fresh deposits could be located with electro-magnetic detectors on low-flying helicopters.

Such a breakthrough, if it occurred, could have immense implications because most gold deposits so far known have been located by some surface outcrop (even the deep South African reefs). If buried orebodies with no surface clues could be spotted, then a new dimension in gold production could be opened up. The significant point to realise is that only very recently has the full weight of fresh scientific and technical know-how shifted to gold. We are witnes-

3 *Nature*, vol. 321, 1986, pp. 854-61

sing the biggest revolution in gold mining for a century, and it is still early days. No one can predict where it may lead, especially if the price stays over $400 in terms of 1987 dollars. Most of the industry's expansion in this decade has been in or around previously known deposits. The next stage will be to make finds in virgin territory.

Yet, already, gold mining analysts are having to revise upwards every year their estimates of future production. Conventional wisdom in 1980 was that Western gold mine output, excluding South Africa, would stay around 300-350 tonnes a year; it is already double that. Predictions that production would level out after 1987 came to nothing. Output is set to increase at least until 1990. Thereafter it may pause slightly when many open pits with short lives will be worked out, and new mines will increasingly just replace old. Environmental controls may also bite harder, making some projects uneconomic. The environmental lobby is already strong in some American and Australian states, as we shall see in the following chapters. Another factor which may slow down growth is that many of the best long-term prospects are in countries like Indonesia, Papua New Guinea, the Philippines or West Africa where the local infrastructure is poor, so costs are higher. Ultimately, it is all a matter of the gold price.

Undoubtedly there is plenty of gold to be won. Everywhere I have travelled to look at new gold prospects, in Australia, Brazil, Canada, China, Indonesia, the United States and Venezuela, I found many genuine new prospects that were not even included in most analysts' tables of future production. Such a renaissance of gold mining has dawned that forecasts are outdated almost overnight. To say, in the light of what we know today, that Western gold production may level out by 1990, giving a plateau of around 1450-1550 tonnes annually through the next decade, sounds reasonable. So does the proposition that Chinese and Soviet output could rise to about 500 tonnes. But those statistics could easily be overtaken by events and, in the light of what one would have said just five years ago, probably will be. The essential point is that we are in a new era for gold mining. As one mining consultant put it, 'An understanding of the forces at work is of greater importance than any formal forecast.'

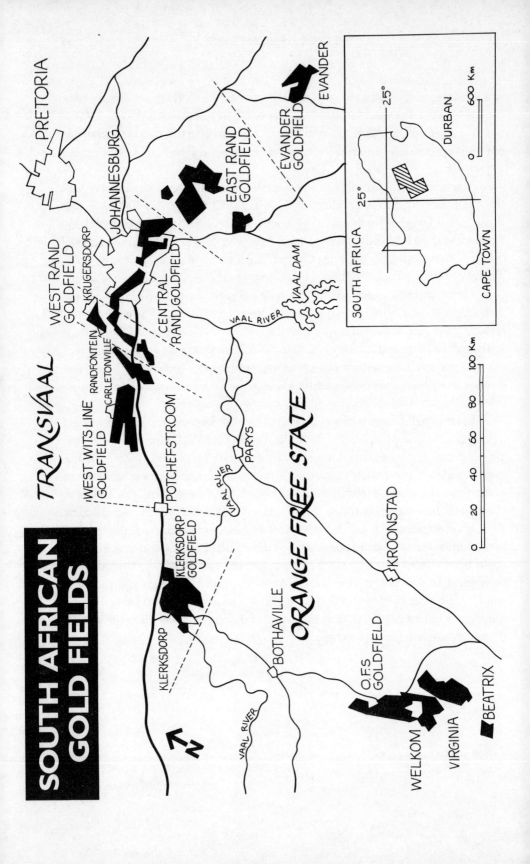

SOUTH AFRICAN GOLD FIELDS

TRANSVAAL

PRETORIA

JOHANNESBURG

WEST RAND GOLDFIELD

KRUGERSDORP

CENTRAL RAND GOLDFIELD

EAST RAND GOLDFIELD

EVANDER GOLDFIELD

EVANDER

VAAL DAM

WEST WITS LINE GOLDFIELD

RANDFONTEIN

CARLETONVILLE

VAAL RIVER

POTCHEFSTROOM

KLERKSDORP GOLDFIELD

KLERKSDORP

VAAL RIVER

PARYS

BOTHAVILLE

ORANGE FREE STATE

KROONSTAD

O.F.S. GOLDFIELD

WELKOM

VIRGINIA

BEATRIX

VAAL RIVER

N

0 20 40 60 80 100 Km

SOUTH AFRICA

25°

25°

DURBAN

CAPE TOWN

0 600 Km

Chapter 2

SOUTH AFRICA:
THE POLITICS OF GOLD

In the autumn of 1986, South Africa celebrated the centenary of Johannesburg as the home of the world's foremost gold mining industry. In the hundred years since gold mining took off in the Wiwatersrand, over 40,000 tonnes of gold have been mined; about forty per cent of all the metal produced in history. South Africa has made a unique contribution to the world of gold. She has stood head and shoulders above any other producer, quite in a class of her own, yielding, until the 1980s, well over seventy per cent of Western gold output, and over half of world supplies.

The centenary should, indeed, have been a celebration. Instead it was a curiously muted affair. Although over 600 delegates turned up for the Gold 100 conference in Johannesburg, there was not a single member of the London gold market. Yet it was the London market that had marketed most of South Africa's production for a hundred years, and it was London refineries that handled all the gold until the Rand Refinery opened in the 1920s. No American gold traders showed up either. The event underlined the problems and the isolation now facing South Africa.

Strictly on the mining front, the gold business there moves into its second century in fine shape. As one Johannesburg analyst put it, 'Profit margins are at near record levels, exploration and expansion activity have never been more intense. The industry continues to create new jobs faster and in greater numbers than in any other industry in South Africa'. On the great gold fields of the West Wits Line and the Orange Free State, new 'supermines' are being established by the merger of old ones. The real giant is Anglo-American's Freegold, created from the combination of five existing mines, which will yield up to 110 tonnes of gold annually by the early 1990s (a single mine on a par with all of Australia's or Canada's output). That is only the beginning of the story. Drilling rigs are mushrooming on the veld beyond the West Wits Line, and to the south of the Orange Free State, as the mining houses continue their geological detective work to sniff out extensions to the deep gold-bearing reefs on which South Africa's fortunes have been founded. The word is that by the early 1990s, three, maybe even five, new mines are in prospect beyond West Wits and a further three or four in the

Orange Free State. Even conservative predictions by the Chamber of Mines suggest South African production staying around 700 tonnes annually for the rest of this century. Those drilling rigs may be signposts to even more.

The NUM Makes its Stand

The gold is certainly there, and the mining houses have the know-how to extract it (even down four kilometres deep). But what is the political future of South Africa to the end of this century? That is the crux. The mining industry has to face an increasingly uncertain and violent political horizon. Within its own domain, the black National Union of Mineworkers (NUM), led by an engaging and impressive former lawyer, Cyril Ramaphosa, is hard at work recruiting the 460,000 Africans who comprise over ninety per cent of the work force on the mines. The Union reckons to have nearly 360,000 members, and is slowly becoming a more organised and potent force in everything from wage negotiations to safety. (The Union's stance on safety was stiffened by the disaster at Kinross gold mine in September 1986 in which 177 miners, nearly all black, were killed.) The growing economic and political aspirations of this huge black work force, more aware of their collective strength, cannot be ignored by the mining houses. In the past their attitude has often been to sack anyone who went on strike, confident that plenty of other Africans were waiting to take his job. The new muscle of the NUM, however, was demonstrated when miners, sacked after an official strike called by the Union at the Marievale mine, were reinstated on the orders of the South African Supreme Court. The Court ruled that the NUM had followed the correct disputes procedures, and the mining house, Gencor, had exceeded its authority.

The decision was a valuable milestone in establishing the NUM's members' rights. Even so, the NUM does not have an easy task. The mines still rely essentially on a migratory work force, so there is a constant ebb and flow of men coming to take up new contracts of nine months to two years, while others return home. Nowadays the majority of these workers come from the so-called 'homelands' of South Africa and Lesotho, whereas traditionally they came from other African nations such as Botswana, Malawi and Mozambique. Today scarcely twenty per cent of the migrant workers come from abroad. But unemployment is endemic in the 'homelands' and in Lesotho, so a job on the mines remains a magnet and their local economies can barely survive without the remittances sent home by the miners. Consequently it is hard to build a permanent Union membership with the commitment to organise and sustain a long strike that could cut back production severely. The thirty-five gold mines, after all, are spread out in a wide arc of six gold fields several hundred miles apart. To keep a strike going on all of them for a lengthy period, when faced with tough mine security guards armed with tear gas, plastic bullets and dogs, to say nothing of the ruthless South African police and army, is a major battle. Moreover, most of the mines maintain three or four months of ore reserves on the surface, which would enable them to keep up some level of gold output even if underground work was halted. The chances of a serious cutback in gold production are thus limited, and the effect on the gold price when disruptions do occur is likely to be psychological, rather than real. Gold is much less vulnerable than platinum, of which South Africa supplies over eighty per cent of Western production from just two main mines. Halting two platinum mines is much easier than crippling thirty-five gold mines.

Yet the growth of the National Union of Mineworkers has opened a new era for South African gold mines. The Union has a constructive job to do in improving wages, living conditions and safety. Cyril Ramaphosa is determined, in particular, to hasten the end of the migrant labour system, seeking to replace it with a more stable, better-paid and skilled work force able to live at the mines with their families. The present system of hostels organised on tribal lines inevitably creates a sense of restlessness and tension, which has erupted in much bitter inter-tribal fighting that has cost many lives. At just two mines; Vaal Reef and President Steyn, over eighty men died in tribal clashes in late 1986 and early 1987. Anxious to break the spiral of violence, the NUM voted at its 1987 Congress to dismantle the migrant system 'within a reasonable time'. Ramaphosa sees that as one or two years. He is pressing, too, for even better wages. The Union got 23.4 per cent in 1986 and initially asked for 55 per cent in 1987. Although wages are up over ten-fold in real terms since 1970, the gap between earnings of white and black miners is still wide. The white Mine Workers Union, with only 15,000 members, has fought a long and bitter rearguard action to prevent black advancement to skilled jobs, and, above all, to the ultimate accolade of a 'blasting ticket' to supervise explosives underground. Breaking down that barrier is crucial to the NUM's advancement. In a word, as Cyril Ramaphosa says, 'Miners can only reach the highest wage levels if the apartheid system is completely eradicated.'

Thus the development of the NUM is inextricably mixed with the political future of South Africa. Gold provides close to fifty per cent of South Africa's foreign exchange earnings, and it has been, in a phrase used so often, the 'flywheel' for expansion not just in South Africa itself but throughout southern Africa for the last century. The strength of South Africa is also its Achilles heel. If gold production (or gold sales) *could* be halted, then the minority white government would be much less able to fly in the face of an increasingly hostile international community. Inevitably the NUM has become a conduit for black political aspirations. The focus will shift more and more on their ability to determine the destiny of the industry. As Mrs Winnie Mandela, wife of the jailed African National Congress leader Nelson Mandela, told miners mourning the loss of their colleagues after the Kinross disaster, the worst in the industry's entire history, 'You hold the golden key to our liberation. The moment you stop digging gold and diamonds, that is the moment you will be free.'

Can the Mining Houses Respond?

The mining finance houses, naturally, are anxious to dodge the storm clouds. The greatest of them, Anglo American, whose five main gold mines account for over 250 tonnes of gold (nearly forty per cent of all output) has long been in the forefront of reform. Anglo American, built up initially by Sir Ernest Oppenheimer, and then run from the mid-1950s to the mid-1980s by his son Harry Oppenheimer, has always been more generous in wage awards, in improving conditions on its mines, in helping black education and has never hesitated to criticise the government. Harry Oppenheimer was a constant advocate of political reform, and Anglo's sister company, De Beers, astride the world of diamonds, works closely with many governments throughout Africa.

Oppenheimer's successor, Gavin Relly, has equally pressed for the abandonment of apartheid, and tries to foreshadow what post-apartheid South Africa will be like. 'We are making a determined effort to visualise what credible will mean ten or fifteen years from now,' he wrote in *Foreign Policy* magazine. 'It will certainly not mean a simple projection of current manning and industrial relations practices, however progressive they may be. Credibility will require a mind-wrenching effort to grasp the future and translate it back to the real action we should be taking now.' At the outset, he urged engaging in much closer dialogue with black political leaders about the future, and set the tone himself by leading a delegation to meet in Zambia with the banned African National Congress.

Mining houses like Anglo do not have the tunnel vision of the Afrikaaners, who dominate the government and civil service of South Africa. Equally they have a great deal to protect, and must anticipate dealing with a black majority government one day. Undoubtedly, whoever is in power in South Africa is going to need the gold mines as the prime export earner.

Ultimately the National Union of Mineworkers is committed to the nationalisation of the mines, but that may become a less dogmatic goal. Both the NUM and the exiled African National Congress (ANC) realise South Africa's wealth is based on gold and, pragmatically, they do not wish to jeopardise that asset, provided the miners themselves are sharing the wealth and can live and work at the mines in greater dignity. The ANC is shifting to a position which 'ensures that the wealth of the country increases significantly and continuously'.

This brings us back to the narrower theme of the prospect for the gold mines themselves. Right from the start, the gold mines of South Africa have been a breed of their own. The very nature of the gold, finely disseminated in reefs plunging more than 4,000 metres into the earth, meant that it could not be mined by the diggers who prospered in the California or Australian gold rushes. It called for capital on a grand scale, and exceptional geological and mining skills. Today it requires a capital investment of anything from $500 million upwards and takes seven or eight years to bring a new mine into production. Thus the mining finance houses grew up as the nucleus which could raise vast amounts of money to nourish teams of geologists and mining engineers.

The original cast of six houses has not changed. Number one is Anglo American, pioneered by the Oppenheimers, whose particular trump card for many years has been the Orange Free State gold field, together with such nuggets as Western Deep Levels on the West Wits Line, the world's deepest gold mine.[1] Their nearest rivals are Gold Fields of South Africa (GFSA), originally founded by Cecil Rhodes in 1887, and now steered on a very tight helm by Robin Plumbridge, who views himself and his team as miners who are not to be distracted by political excursions or seduced by the lures of forward gold sales, or futures or options markets. Gold Fields produces about 130 tonnes of gold a year (twenty per cent of South African output) and sells it at the going market rate through the Reserve Bank of South Africa (as all mines must by law). It is fortunate to have under its control the lowest-cost mines, which is one reason why the dazzling lights of the futures/options casino have less attraction. Hedging is less essential than on high-cost marginal mines. Operating costs at GFSA mines like Driefontein Consolidated and Kloof are still scarcely $110 an ounce. Gold Fields mines also all have long lives, and with one exception, Venterspost, are set for the twenty-first century. Although GFSA is an independent mining house in its own right, Consolidated Gold Fields in London does have a forty-eight per cent shareholding.

Challenging Gold Fields for second position is the Gencor Group (a merger of two old houses, General Mining and Union Corporation) with a clutch of fourteen rather diverse mines, mainly at the Evander gold field and in the Orange Free State, which produce about 130 tonnes a year. Gencor has grown up with much closer Afrikaaner connections in finance and management than the other houses. Consequently it has always cut a rather stern, conservative image and been notably tough in wage bargaining and union reception. However, it now has Derek Keys, an English-speaking outsider, as chief executive, and is seeking to polish its image.

The three smaller houses in gold are Johannesburg Consolidated Investments (JCI), with three mines (including the newcomer Joel) which give it just over 60 tonnes annually; Barlow Rand, with five mines worth 70 tonnes annually; and Anglo Vaal, with three main mines notching up about 40 tonnes.

1 The individual mines of each house are listed in appendix IV

Among these six houses, however, there is the extraordinary cross-fertilisation of share holdings that is unique to South Africa. The mining houses not only hold large blocks of shares in their rivals, but often have at least one seat on the board. Although one house will manage the individual mines under its umbrella, the others will also usually have shares. Competition in South Africa is really limited to the detective work by the geologists in finding and proving up a new mine; once it is floated, everyone gets a slice of the action. Anglo American, in particular, has woven its web into most of the other houses. It has a 41.3 per cent stake in JCI (with De Beers, Anglo's diamond associate, controlling another 9.1 per cent), 10.8 per cent in GFSA, and smaller holdings in both Barlow Rand and Gencor. For good measure, the Anglo-De Beers conglomerate also has a 28.9 per cent slice of Consolidated Gold Fields in London through its Bermuda-based holding company, Minorco. Thus Anglo actually has seats on the board of houses controlling well over seventy per cent of South African production.

Given the size of this empire, it is not surprising to find Anglo, not just the forerunner in political initiatives, but in shaping how the mining industry will evolve to the end of this century.

Two strategies have emerged to ensure a high level of gold production beyond the year 2000. First, the existing mines are being expanded through new shafts plunging to even greater depth, taking advantage of new technological discoveries and much greater mechanisation. Secondly, exploration budgets have been stepped up to pinpoint entirely new gold fields beyond the fringes of the existing ones. The investment in these twin strategies, which is now absorbing close to $1 billion annually, is a measure of the confidence of the mining industry. Despite the lack of new foreign investment in South Africa, the mining houses are pressing ahead to expand their industry as if the political hurdles did not exist. Of course, it cuts both ways. The potential earnings from gold become an even greater lifeline to sustain a beleaguered country. The concept that gold is a lifebelt for all seasons was never truer than for South Africa.

The activity, however, is remarkable compared to the mood in the mid-1960s when I first went to South Africa, and found the gold miners dispirited by the long run of $35 gold. The talk then was of the decline of gold production, almost its extinction, by the end of this century. As it happens, production has fallen from a peak of just over 1,000 tonnes in the 1970s to scarcely 640 tonnes in 1986, but this has much to do with the mining of lower-grade ore. Back in 1970, when gold was still stuck around $35, the average grade mined was 13 grams per tonne; it is now down to 6 grams. But the re-establishment of the gold price on realistic levels has given the industry a new lease of life, which has only come to be appreciated in the late 1980s. The talk now is of pushing production back up over 700 tonnes, and even well beyond. While the high gold price in dollars in the

1980s was an incentive, the declining rand against the dollar maintained much of that benefit. In rand terms, the gold price was higher in 1986 than it was in 1980; operating costs for the industry averaged only around $200. In short, despite inflation and much higher wages paid to the black work force, the industry is very profitable. Those profits have encouraged major expansions.

Burrowing to 4,000 Metres

The best example of the unique character of South African gold mining is at Anglo's Western Deep levels. This mine in the Far West Rand field first opened in 1962, burrowing down initially to the Ventersdorp Contact Reef and then on to the rich Carbon Leader reef far below at 3,000 metres. The original concept of Western Deep was bold enough; it was envisaged to have a sixty-year life, during which it would yield over 2,000 tonnes of gold. The two major problems to begin with were heat (the rocks are over 50°C at such depth) and great underground reservoirs, which meant pumping out 100 million litres of water daily from each shaft. These obstacles were overcome and Western Deep settled into regular production of almost 40 tonnes annually. The Carbon Leader Reef, however, showed no signs of petering out at the 3,000 metres originally planned for the mine. Anglo determined to follow it. In 1980 they embarked on the biggest-ever capital expansion programme for a South African mine, investing over 700 million rand (then over $700 million) to step up Western Deep's production by over seventy per cent before the end of the century. Initially they put in a new 2,734-metre shaft down to the Carbon Leader, which was completed in 1986. A sub-shaft beyond this towards 4,000 metres should be ready in 1992, taking miners deeper into the earth than man has ever gone. Once the new areas of reef can be fully worked in the mid-1990s, Western Deep's output is forecast to peak by the end of 1998 at almost 65 tonnes, an overall increase of seventy-three per cent.

This new era at Western Deep has been made possible by many technological advances, which are now transforming the whole shape of the gold mining industry in South Africa. Traditionally, because of low labour costs, mines were labour-intensive. No real push took place until the late 1960s to develop new equipment to undertake the difficult task of extracting a very thin seam of gold, often less than a metre thick, from a reef 2-3,000 metres down. The mines were trapped into a regular twenty-four-hour cycle of drilling, blasting and clearing up. A great deal of waste ore also had to be hauled to the surface.

Today new rock cutting and hydraulic drilling equipment developed by the Chamber of Mines Research Organisation is gradually phasing out the use of explosives. This will make continuous mining possible (previously everything had to halt for the daily blasting), and reduce the need to haul so much waste to the surface. The waste, instead, can be used to 'backfill' the stopes from which the gold seam has been extracted, so improving underground safety. This more selective mining will also cut the need for so many unskilled workers underground, improving productivity substantially. One simple benefit of fewer underground workers is that the hoists can be used more efficiently to bring out ore instead of people. More important, a smaller skilled work force will be able to operate at much greater depths.

The environmental problems of maintaining efficient working of a large unskilled work force in temperatures above 55°C encountered at depths towards 4,000 metres, had previously been uneconomic. Keeping smaller teams cool should prove easier. Already, at the Harmony gold mine in the Orange Free State, one of the world's biggest ice making plants turns out 20,000 tonnes of crushed ice daily, which is pumped through steel pipes along the mine tunnels. The ice absorbs tunnel heat, melts and the warm water is pumped back to the surface. Such cooling plants on South African mines now have a capacity equal to nearly 3.5 million domestic refrigerators.

The design and layouts of shafts and tunnels at great depths is also being worked out with computer-simulation techniques to try to avoid rock bursts, which cause many casualties in the deep mines. The pressure 2-3,000 metres down is so intense that rocks literally explode.

New technology will thus enable the industry to move into the 1990s with more mechanisation, higher productivity, a leaner work force and, one hopes after the Kinross disaster, a better safety record.

Technical advancement, however, is only one part of the story. The other has been consolidation of existing mines with extensions or new mines alongside, turning them into 'supermines'. The advantages are not just the economies of size, but the resulting tax breaks. If a highly profitable mine plans a major extension or has the prospect of a new mine next door, the sensible thing is to merge them and write development costs off against profits. The taxman loses in the short term, but the argument that has made him agree is that in the long run this ensures the survival of the industry well into the next century. The taxman may not get a huge cake today, but he'll get little cakes much longer. The recipe has been swallowed.

Initially, Gold Fields of South Africa merged its West Driefontein and East Driefontein mines before tacking on a neighbouring property, North Driefontein, that was destined as a new mine. The resulting supermine, Driefontein Consolidated, produces close to 60 tonnes a year, and is set to maintain that level for the rest of the century.

Likewise, Anglo American knitted together its mines on the Klerkdorp gold field into the Vaal Reefs complex, which will also produce between 80-85 tonnes a year throughout the 1990s.

That was almost a trial run for the ultimate merger. Anglo American announced in December 1984 that they planned to bring together all five of their Orange Free State mines into a conglomerate called Freegold. This giant embraces Free State Geduld, President Brand, President Steyn, Welkom and Western Holdings, whose combined output is between 100-110 tonnes a year. Freegold thus accounts for over eight per cent of all Western gold mine production, and its annual yield is worth nearly $1.5 billion with gold at $400. The main argument for Freegold was that the assembled mines were past their peak years of production, and grades had fallen to around 6 grams per tonne. Given the economies of large-scale working, they can soldier on united until well into the next century; as single mines, they might not have survived. Moreover, Anglo does have up its sleeve the long-dormant Jeanette mine just north of Freegold. Jeanette was the group's one failure in the Orange Free State field, and it has been closed since 1955. It might just be re-opened under the Freegold umbrella.

New Mines on the Horizon

Anglo American's attention, however, may well be diverted by more profitable possibilities just to the south of the Freegold mines. The way ahead has already been pinpointed by Gencor's new Beatrix mine a few miles south of the main gold field, which opened in 1984 and will reach peak production of 13 tonnes annually by 1988. Nearby, Johannesburg Consolidated Investments is bringing in the Joel mine in 1987. This small mine, named after one of the founders of JCI, W.J. Joel, is planned to turn up 6 tonnes a year well beyond the year 2000. Anglo, JCI and Gencor are all busy drilling other properties throughout the area, focussing primarily on the aptly named Eldorado reef, tapped already by the new Beatrix mine. Mining analysts, who love to spend their days scouting the location of new drilling rigs, forecast that at least three, and perhaps four, new gold mines may come out of all this activity. No one will speculate how much gold they might yield, but South African mining houses hate mini-mines, and for them that means less than 5 tonnes annually. Ideally they prefer 10-15 tonnes annually with a life of at least thirty years before they will take the plunge. So if four new mines opened in the southern Free State, they might add 30, perhaps even 50, tonnes to South African output in the late 1990s.

Exploration in 'the surging south', as one Johannesburg analyst dubs it, is more than matched by what he also called 'West Side Story', or 'stretching the West Wits Line'. The West Wits Line, more formally known as the Far West Rand Goldfield, was originally discovered in the 1930s by a young German geologist, Rudolf Krahmann, using a magnetometer to chart the pattern of magnetic shales and then plot the location of the associated gold reefs. Krahmann detected not only the Main Reef, on which South Africa's fortunes were originally founded, but discovered too the Ventersdorp Contact and the Carbon Leader reefs. His success transformed the fortunes of Gold Fields of South Africa, which backed him with a modest trial fee of $830 and $560 expenses. GFSA's Driefontein Consolidated and Kloof are on the West Wits Line, as is Anglo's Western Deep Levels.

The intriguing question has always remained how far does the Line extend? Does it swing round towards the next major gold field, Klerksdorp, fifty-five kilometres to the south-west? Both GFSA and Anglo American are convinced it does. Out near the little town of Potchefstroom on the Mooi river midway between the two gold fields, they have a forest of drilling rigs boring down in a kidney-shaped area of 20,000 hectares. The ground staked out is not much smaller than either of the other two big fields. The borehole results are a closely guarded secret, but gossip suggests that some have yielded up to 125 grams (or nearly 4 ounces) per tonne, an astonishing grade compared to the current South African average of 6 grams. A few bore holes do not make a mine. No fast decisions are expected. Neither Gold Fields nor Anglo is likely to jump before 1989 or 1990, when they have had ample time to consider a wide range of borehole samples. But a real possibility of two or three new mines exists. They could add another 30-50 tonnes to South African output. On top of all this there is the Evander goldfield, developed to the east of Johannesburg by Gencor, where the new Poplar mine started up in 1987 and will yield 10 tonnes annually by the 1990s.

What does it all add up to? Ultimately much will depend on the gold price itself. A high price will enable many of the older mines to keep going for the rest of the century; a low one might force their closure, so that new mines coming on stream would only compensate for their loss. Certainly a high plateau will be maintained. One conservative analyst shows production rising initially to 700 tonnes by 1990 and then to 720 tonnes by 1995. He estimates at least twenty-four of the present thirty-five mines will still be going in the year 2000. He includes in that assessment all the 'supermines' like Freegold, Driefontein and Vaals Reefs. But he has not allowed for decisions to go ahead with new Free State and West Wits mines, which could between them yield another 60-80 tonnes. The implication is that South African production could not only rise above 700 tonnes, but it might just go towards 750 tonnes during the 1990s if new mines are proved, and if the price encourages both their development and longer life for some old-stagers.

The price, of course, works in different ways. By law, the mines must extract the lowest payable grade (to preserve a long life for the industry). So a few years of high prices could limit production because of lower grades, but would mean more gold in the long haul. Output was down both in 1985 and 1986 precisely because a high rand price initiated lower grades, and a strong price could check a rise in tonnage for several more years.

Mini-mines – the Hopeful Signposts?

However, a fresh dimension in South African gold mining will notch up extra production. Several entrepreneurs have set about outside the framework of the six mining houses and the Chamber of Mines to revive some old mines and tailings dumps. Joe Berardo with his Egoli group, Loucas Pouroulis with Golden Dumps, Adolf Lundin who took over the old Daggerfontein mine from Anglo American, Les Holmes with Sub Nigel, and Steen Severin with Rand Leases have all shown that small-scale gold operations can be profitable. 'The entrepreneurial spirit with low overheads can do things the giant houses cannot afford,' Richard Johnson of Golden Dumps explained. Golden Dumps, operating from a small bungalow surrounded by flowering shrubs in a quiet suburb of Johannesburg, started by cleaning up tailings dumps and plants at abandoned gold mines. Then they decided to open up an old Gencor mine, South Roodepoort on the Main Reef. Gencor had let the plant run down completely, and Golden Dumps had a tough time trying to bring the old mine and workings back up to scratch. Eventually they did some drilling of their own on the mine reef, found some good values on the Kimberley Reef and started up their own virgin mine on the property. It now produces 1.4 tonnes a year.

Another newcomer, Steen Severin, started out heap-leaching the old sandy tailings dumps at Anglo Vaal's Rand Leases mine. He managed to extract 200 kilos a year from the sand, which he sold to building contractors. But he realised he was working above the dormant mine itself, which Anglo Vaal had closed in 1972. He also found that Anglo Vaal owned only five per cent of Rand Leases, so he quietly began buying up the shares, eventually securing twenty-five per cent. That gave him effective control. So he did a feasibility study and decided that for a modest 33 million rand (about $16 million) he could bring the mine back into production. Anglo Vaal, by contrast, had estimated it would cost about $50 million to get going again. They did not consider, Severin told me with some delight, the advantages of carbon-in-pulp for extracting the gold in their calculations. He did. So Rand Leases started producing gold again, after a sixteen-year lapse, early in 1987, and should produce 1.8 tonnes a year. Severin says the advantage he has over the big mining houses is simple: 'It's my own business. I'm looking for ideas day and night.' His wife works with him, seeking to establish good rapport with their growing African work force. 'Excuse me,' she said, halfway through our talk. 'I have a meeting with a hundred of our Africans.'

The entrepreneurs have been successful in recruiting promising young geologists and mining engineers. 'It's nice to get away from the big mining house syndrome,' one engineer, who had forsaken Rand Mines, told me. 'There's lots of small mining opportunities.' In reality, the actual production from these newcomers is limited, but they could add another 10-15 tonnes to South African output in the early 1990s.

Perhaps, more significantly, the small operators have greater flexibility in labour relations. In some instances, they are already paving the way for an open multi-racial workforce out of the spotlight that is inevitably on the main houses. 'We are trying to build small teams of two or three to work together underground,' said Richard Johnson of Golden Dumps, 'with each team developing their own stope. And we don't mind whether people are black or white. Eventually all our underground miners will be in a team, with no regard for their colour. Already we've got whites driving trucks and shovels underground side by side with blacks for the same pay.'

Such equality is rare in South Africa, but can it foreshadow the future? These small mines could provide a much-needed testing ground. For the dichotomy is that the industry itself is in good shape in terms of exploration, expansion and new technology. Plenty of gold is there to be mined. The question is, will the political climate allow it?

UNITED STATES GOLD MINES

OREGON | IDAHO

NEVADA

SLEEPER ▲

CHIMNEY ▲
▲ JERRITT CANYON

PREBLE ▲ ▲ GETCHELL
▲ PINSON ▲ BOULDER CREEK
CARLIN ▲

SALT LAKE
CITY →

80

BATTLE ▲
MOUNTAIN RAIN ▲

MAGGIE CREEK
GOLD QUARRY

BINGHAM

▲ HORSE CANYON
▲ BUCKHORN

MERCUR ▲

RELIEF CANYON ▲

EUREKA ▲

RENO ○

▲ BALD
MOUNTAIN
ALLIGATOR RIDGE

50 ▲

EUREKA ▲ ▲

ELY ○

COMSTOCK ▲

WINDFALL ▲ TAYLOR

UTAH

▲ RAWHIDE

NORTHUMBERLAND
▲ ROUND MOUNTAIN

BOREALIS ▲

▲ MANHATTAN

CALIFORNIA

▲ GOLDFIELD

U.S.A

95

WASHINGTON

ZORTMAN LANDUSKY ▲

OREGON

▲ MONTANA TUNNELS

WEST END ▲
IDAHO

HOME
STAKE

WYOMING

LAS VEGAS ○

HAYDEN
HILL ▲
YUBA
PLACER ▲

McLAUGHAN ▲

MERCUR ▲
UTAH

COLORADO
SUMMITVILLE ▲

ARIZONA

JAMESTOWN ▲

SUNNYSIDE

CALIFORNIA

ARIZONA

NEW MEXICO

PACIFIC OCEAN

MESQUITE ▲ ▲ PICACHO

Chapter 3

THE UNITED STATES:
GO WEST, GOLD MINER

When I first studied American gold mining in the 1960s, it was a depressing scene. Long gone were the heady days of the California gold rush and the surge in output in the late 1930s, prompted by gold's initial rise to $35 an ounce, had petered out. The only serious gold mine remaining was Homestake at Lead, South Dakota, opened in 1877 and still going strong, plus Kennecott's Bingham Canyon copper pit in Utah, which yielded plenty of by-product gold. The sole encouraging news was a mine just opened by Newmont at Carlin in Nevada, which held out promise from an orebody liberally sprinkled with finely disseminated particles of gold. No one realised, however, what a significant signpost to the future Carlin then presented. Anyway, few miners were really interested in looking for gold while the price stuck at $35. 'There's a great deal of gold in the ground,' J. Patrick Ryan, the gold specialist at the US Bureau of Mines told me in 1967, 'it's all a question of the economics of getting it out.'

Those economics were transformed only in the late 1970s, when North American mining companies adjusted to the notion that higher real prices for gold had come to stay. At last they pulled out the old geological surveys to see what might have been overlooked as unprofitable in the past. Homestake, the doyen of American gold mining houses, noted they had explored a possible hot springs gold system at Cherry Hill, to the east of California's Napa Valley, in 1926. They went back for another look in the neighbourhood and found a large disseminated orebody, grading just over 5 grams per tonne. On 2 March 1985 they poured the first bar of gold at the McLaughlin mine, which is set to produce 5-6 tonnes of gold annually well into the next century.

By then, Homestake were no longer alone in the field. As base metal (and later oil) prices fell, a host of North American mining groups diverted all their energies into the search for gold, especially in California and Nevada. American output responded. After stagnating at around 30 tonnes a year for most of the 1970s, it surged to about 130 tonnes by 1987, and is projected to reach at least 160 tonnes, perhaps even 175 tonnes, by the end of this decade. The sheer task of keeping track of new projects has proved hard. The US Bureau of Mines admits they have difficulty totting up production figures and are constantly revising their statistics. At least twenty-eight mining groups are working on forty-five gold mines, either in production or scheduled by 1988. Newmont Gold, the spin-off from Newmont Mining, has edged out Homestake as the largest producing group. Other newcomers are making their mark. American Barrick, Battle Mountain Gold (a spin-off from Pennzoil), Echo Bay Mines, Gold Fields Mining Corporation (the American arm of Consolidated Gold Fields), Freeport Gold (the gold spin-off from Freeport McMoran) and Pegasus Gold are all contributing significant tonnages.

The boom has been achieved without anyone really pinpointing a new gold field. 'It's the history of mining,' said John Lucas, now the gold specialist at the Bureau of Mines. 'Mines all have nine lives and you zero in on old ones, re-examine old ores and with new ideas and new technology come up with a hot item.' John Lucas believes the momentum will be maintained for quite a while: 'The geologists will remember features from one to identify another,' he went on.

My own travels have certainly convinced me that America's new gold rush has plenty of life. A high level of production will be maintained for the rest of this century. Exploration is spread throughout the American West, from Montana and Washington down through California and across Utah and Colorado, but the pacesetter is Nevada.

Nevada, Gambling on Gold

Nevada has long called itself 'The Silver State', a tag originating from the rich Comstock Lode near Virginia City which yielded almost $200 million in silver in the 1860s and '70s. Today it might well be re-christened 'The Gold State'. Close to thirty new gold mines have been opened in the last few years, and the hunt for more orebodies continues apace. Motel rooms are hard to come by in the little towns of Elko and Winnemucka on Interstate Route 80 that straddles from east to west across the dry, hilly country of northern Nevada, for they are mostly taken up by geologists and exploration crews. According to one account, ninety per cent of all mining geologists in the United States are now working in Nevada. At 4.00 am every morning, just as gamblers of a different kind are going to bed, they may be seen packing their equipment into dusty pick-up trucks and heading for the hills.

'The number of claims staked is spectacular,' one geologist told me. 'Even if the road looks pretty they'll stake it, and they run into each other in the field and follow each other round.' Geologists for the major companies like Newmont, Gold Fields or Freeport are tracked most diligently, on the assumption that with big exploration budgets and back-up expertise to call on, they must have something up their sleeves. Often they have. Gold Fields turned up a prospect called Chimney in the Osgood Hills near Winnemucka that conservatively contains 140 tonnes (4.5 million ounces). 'It's just beginning,' said John Schilling, director of the Nevada Bureau of Mines, 'if we had a slightly higher price you could probably mine half the state. As it is, ninety per cent of Nevada's mines break even at $200, make money at $250, and a damn good profit at $350.'

Small wonder, then, that with such margins the hunt is on. Nevada's output has already quadrupled from 8.6 tonnes in 1980 to over 35 tonnes in 1986 and is forecast to top 70 tonnes (2.2 million ounces) by 1990, accounting for half of all US gold output.

The difference, however, from the bonanza at the Comstock Lode a century ago is that in place of high-grade lode mines, the newcomers are all very low-grade epithermal deposits of disseminated gold - micro-sized particles quite invisible to the naked eye - on or just below the surface, that can be mined as open pits. The grade is usually 3-5 grams of gold per tonne, but some open pits get by on scarcely 1 gram, if they do not have to remove too much topsoil and waste ('overburden', as the professionals call it) and can heap-leach. Nevada's tolerance of open-pit mining helps. 'Nevada has the best climate for mining of any state in the Union,' one mining executive explained. 'The paperwork and permits are much less. Half these projects would never start up in California, with all its environmental controls.'

The first signpost to this era in Nevada was planted in the early 1960s through some smart geological detective work by geologists John S. Livermore and Robert B. Fulton for Newmont Mining. Scouting the Tuscarora Mountains beyond the Humboldt River north-west of Elko they came across a relatively high-grade deposit, reaching 11 grams per tonne, of microscopic gold particles - so tiny they had to be magnified 1,800 times before they could be photographed - that had been entirely overlooked by Chinese gold diggers, who operated lode mines in the same hills at the turn of the century. The orebody was right on the surface and could be mined as an open pit. So Newmont started the Carlin gold mine in April 1985, ushering in a new age of gold in the American West.

The high grade at Carlin meant that it was profitable even with gold pegged at $35, and with gold recovery through the conventional mill. The picture was not really transformed, however, until the late 1970s with gold rising over $200, and the introduction of heap-leaching, carbon-in-pulp recovery and the beginnings of computer control of ore grades. But Carlin pointed the way.

'This is the mother of disseminated gold pits,' said Joe Rota, a geologist with the Newmont Gold Company, as we stood on a bluff more than twenty years later, overlooking a long, narrow canyon nearly a mile long and 600 feet deep from which nearly 100 tonnes of gold had been filleted. Today the original Carlin deposit is virtually mined out. Just a couple of trucks, looking like toys, growled around on clean-up operations. A small herd of deer were already sharing the viewpoint with us and Joe Rota said that the sage brush would grow back in the terraces of this manmade mini-Grand Canyon in a few years. But all around Carlin the hills are alive to the sound of drilling rigs, bulldozers and dump trucks. At noon most days the dull thump of explosives signals fresh 'benches' being cut out in half a dozen other active pits nearby. For within forty miles north-west and south-east along the range of the Tuscarora Mountains, Newmont's geologists have located at least ten similar orebodies, and rivals have staked out others too. What has become known as 'the Carlin trend' is being unveiled. Almost every hillside contains gold. The orebodies are not always as rich as Carlin, but because the basic facilities of offices, assay laboratories, milling plants and heap-leach pads are already in place, many of these 'satellites' can be worked profitably for a few years.

Joe Rota had a detailed map of the terrain. To the north of Carlin he had inked in Bootstrap, Post, Blue Star, North Star, Genesis and Pete deposits; to the south Maggie Creek, Gold Quarry, Rain and Immigrant Spring. 'We've got ten pits, all disseminated gold deposits, but each is unique,' he explained, as we went bouncing along a dirt road in his pick-up truck to view the satellites. 'If we took Carlin as our model, we might miss them. We've learned to take samples of anything unusual. You can't say A plus B equal gold deposit.' The original Carlin pit, for instance, was a high-grade deposit, with the ore thick and continous. By contrast, Gold Quarry, eight miles down the road, which is now the best source at 5 tonnes annually, is a highly erratic orebody. When it was formed by volcanic action millions of years ago, the rocks were less permeable and so 'soaked up' the molten gold less evenly; instead Gold Quarry's rocks are full of rough fractures into which the gold settled. 'Gold Quarry is a different critter to Carlin,' said Joe Rota. On the other hand, Rain, just a few miles farther on, is again a beautiful uniform, continuous orebody that is easy to mine. The precise recipe of each orebody determines how it is treated. At Gold Quarry, for example, ore above 1.4 grams is milled, and lower grade, down to a cut-off of 0.56 grams per tonne, is heap-leached.

Presently we pulled up at Genesis, a small new pit, where we found another geologist, Tyler Shepherd, clambering about a newly blasted 'bench', tapping rocks expertly with a little pointed hammer. He was in a state of some excitement, showing us some samples of newly exposed rock. 'You see the silica, it's just like Gold Quarry,' he said to Joe Rota. 'That's neat, it's high grade.' Pits like this have to earn their keep. If the grade proves disappointing or the orebody fades out too soon, they are abruptly closed down. But the benefit of these open-pit operations is that they can be re-opened again easily if the gold price rises and justifies lower grades, or if another nearby pit yields some very high-grade ore that can be balanced against low-grade to maintain the average target.

Newmont Gold's clutch of pits around Carlin has probably touched the tip of the gold reef. Yet the expansion has already pushed output from a modest 5 tonnes in 1984 to 20 tonnes by 1987, and 28-30 tonnes is forecast for 1990. 'That's a conservative target,' Newmont Gold's manager Ron Zerga told me. 'We are doubling the tonnage of heap-leaching and we may triple it. This is the richest zone for exploration available in North America. It's good at least to the year 2000.'

Newmont's ace is that in 1983 they acquired the whole of the T-Lazy-S Ranch, a wedge of hills and valleys spreading over nearly 900 square miles around the present mines. The 'Carlin trend' runs right through the heart of this terrain but, apart from the narrow strip along the ridge of the Tuscarora Mountains that Newmont Gold has already exploited, the full mineral potential of the ranch is unexplored. 'We haven't found anything new on the T-Lazy-S yet, but we will,' predicted geologist Joe Rota, as we stood on a hilltop overlooking the immense sweep of the ranch across a broad valley to another range of hills beyond.

The possibilities seem endless and not just for Newmont as pioneers. Already, just beyond the perimeter of T-Lazy-S, the Cordex Syndicate's Dee mine is in business, while just over another range of hills a few miles to the north-east the Freeport Gold Company's Jerritt Canyon open pit is steadily producing 8 tonnes a year. Freeport, like Newmont, is happily turning up 'satellites' close to the original find. Moreover, the 'Carlin trend' is no isolated phenomenon. Epithermal disseminated gold deposits pepper northern Nevada. As I drove westwards on Interstate 80, the lonely roads heading off to the horizon from each exit ushered the way to some fresh find. There was Placer Development's little nest of properties in the Cortez Mountains. The original Cortez mine was a pioneer of heap-leaching; now they are working on Horse Canyon and Gold Acres nearby. The next turn-off was to Battle Mountain, once a flourishing copper producer that fell foul of the weak copper price, but has been successfully revived as one of the largest American gold producers in the late 1980s.

Battle Mountain's Fortitude operation can hardly be called a mine or a pit. More accurately, it is slicing away half a mountainside into neat terraces. As you approach up a gravel road through the sage brush from the valley below, there is no sign of life. Then you swing around the corner on a barren hillside, and it is as if a giant had taken a chisel to the backside of the mountain and notched it neatly away in twenty-foot-high steps. The orebody, containing gold, silver and copper is wrapped like a thick overcoat around one side of the mountain ridge. So, starting at the top, the miners simply strip it away. 'The peak is at 7,000 feet, and we've sliced down the benches so far to 6,000 feet,' explained geologist Pat Wotruba. 'It's a fairly high-grade deposit, nearly a quarter ounce (7 grams), with a silver-to-gold ratio of about 5:1.'

Initially the miners had to cut away a great deal of overburden to get at the best of this gold jacket. Their ratio of waste material to profitable ore was 17:1, but by 1987 they had cut into the main orebody and brought it down to a satisfactory 4:1. That reduced costs by $20 an ounce, making Fortitude one of the cheapest producers in North America with cash costs as low as $132 an ounce. Annual output is forecast at 7 tonnes well into the 1990s. As a bonus, a small satellite deposit named Surprise has been located six miles north along the ridge and will be in production by 1988. Battle Mountain's geologists are spending $1 million a year looking for similar satellites in the hills around. 'It's a real hot spot,' said Pat Wotruba, 'we hope for super-economic deposits.'

If Battle Mountain's Fortitude and Surprise deposits foreshadow another complex like Carlin, there is already no doubt about the rich rewards at the next Interstate exit. From the tiny hamlet of Golconda, just a motel, a coffee shop and gas station, a road twists north along the slopes of the Osgood mountains to another blossoming batch of open pits. The first two, Preble and Pinson, controlled by the Cordex Syndicate (which includes Rayrock Resources and American Barrick), are classic examples of small efficient gold pits producing between them a modest 2 tonnes or so a year, by a combination of milling with carbon-in-pulp recovery and heap-leaching. Yet they are highly profitable; at Pinson, the entire cost of the mill was paid off in the first sixteen months of operation. The grade is modest, usually 4-5 grams per tonne, with a cut-off of just over 1 gram, from a cocktail of orebodies. 'We've got five orebodies right here at Pinson,' said Keith Belinghari, the chief mine engineer. One of them was found right next to the milling plant, just covered with a thin layer of topsoil, nestling into the hillside where the edge of the valley butts into the mountain. 'It wasn't even supposed to be there,' said Belinghari. But Mag pit illustrates the continuing surprises in the Osgood Mountains. 'We're spending $600,000 a year on exploration,' he went on, 'we've still got a lot of land under our jurisdiction.'

Pinson is just the forerunner. Freeport's Getchell, just up the road, will soon start up with 4-5 tonnes annually, and Gold Field's Chimney, another sixteen miles up a new dirt road, will contribute as much again by 1989. Moreover, the sheer size of the Chimney deposit, with reserves containing 140 tonnes of gold at grades of 5-6 grams per tonne already billed well in advance of its opening, suggests that the Osgood Mountains hide a 'Carlin trend' of their own.

So just how many such belts of gold are there in Nevada? No one yet knows. The current hunt concentrates on three main strings of deposits in northern Nevada. But there are other rich pickings scattered throughout the state. A couple of hundred miles to the south, Canada's Echo Bay group is beavering away at the Smoky Valley Mine at Round Mountain, which it operates for its minority partners Homestake and Case, Pomeroy. Like Battle Mountain's Fortitude, the operation at Smoky Valley really involves demolishing half a mountain sprinkled with very low-grade gold and heap-leaching it. Circling the site in a small plane with Echo Bay's chairman Bob Calman, it looked like some archaeological excavation of an ancient Aztec city. The whole top of the mountain had been whittled away, and great terraces cut in swatches down the sides.

The huge gold deposit measures a mile and a half across and is 1,700 feet deep. It is being blasted, bulldozed and hauled to heap-leach pads forty feet high at the rate of 18,000 tonnes per day. That alone made Round Mountain the biggest heap-leaching operation in the world when I visited it late in 1986, when output was 5.5 tonnes. Echo Bay's energetic chairman Bob Calman was not content with that. He was planning to double the size of the operation by 1990. 'The key is to have cash costs of about $200 an ounce,' he said. 'We've got 175 million tonnes at that level with a grade of just over 1 gram per tonne. If we can shift 35,000-40,000 tonnes per day, we can get production up to 300,000 ounces (9.3 tonnes) by 1990.' Such an achievement would make Smoky Valley the largest single gold pit in North America.

The intriguing point about Round Mountain, and all those pits along the hill tops of northern Nevada, is that they may indeed be just the tip of the iceberg. An orebody tucked into a hillside is usually right on the surface or only lightly covered by topsoil. Thus it is relatively easy to detect; natural erosion will often reveal part of it. The tougher question facing geologists in the years ahead is what is also out in the valleys hidden by thicker layers of topsoil and offering no clues on the surface? It has become more urgent because Amax Inc. has actually found a small deposit, aptly named Sleeper, in the Slumbering Quinn River Valley due east of the Pinson, Getchell and Chimney mines. The mine has a good grade of 11 grams and promises 1.6 tonnes annually. 'Sleeper is way out in the valley away from the mountain ranges, but luckily was exposed on the surface,' explained a geologist setting out from his motel room at Winnemucka (the nearest town) at dawn hoping to have equal luck, 'but are there a lot more like it in the valleys under three to four feet of soil? That's the real teaser.'

California: Pleasing the Environmentalists

The way to the McLaughlin open-pit mine, set 2,000 feet up in the coastal ranges of California, is pleasant. Our plane landed by a small clear lake, and we drove through graceful apple orchards for a while before winding up through chapparal scrub towards Homestake's new venture. These hills were mined for quicksilver over a century ago, but none of those early miners realised that they were ranging over one of the finest-preserved hot springs gold systems in the world. Yet, according to Homestake's geologist Norman Lehrman, 'The deposit was conspicuous. If you came thinking gold, there it was. An orebody a mile long, 500 feet wide and 400-880 feet deep.' This golden bowl, set between what geologists call two 'hanging walls', contains at least 100 tonnes (3 million ounces) of gold which Homestake plans to mine steadily until the year 2005.

Despite the size of the gold reserve, however, some mining analysts have questioned whether Homestake should have developed the McLaughlin pit. California has tough environmental controls, and they came into play against almost every aspect of McLaughlin. The mine is sited at the junction of three counties, Napa, Yolo and Lake, each with its own bureaucracy and each determined to get the maximum benefit for its own community. Over 260 permits, filling sixteen volumes, had to be obtained. The processing plant had to be sited four and a half miles from the pit itself to be on impermeable ground. The water leaving the site has to be purer than the source stream from which it is drawn. Even the quality of air around the mine site must have lower dust levels than those encountered in downtown San Francisco. The gold ore itself presents another hurdle; it is ninety per cent sulphide and would normally have to be 'roasted' to unlock the gold. But roasting presents additional environment hazards. So Homestake have had to pioneer a pressure oxidation technique, which adds substantially to their costs. All told, Homestake had to invest $280 million to bring the mine into production, and the continuing environmental checks alone add close to $10 an ounce to operating costs. 'We could have brought this mine for $50 million less in Nevada,' a director admitted. Consequently, Homestake is faced with cash operating costs of at least $300, including depreciation, and really needs a price of $400 and beyond to work comfortably. This is double the costs at many of Nevada's new mines.

To their credit, Homestake have not ducked their environmental commitments. The chairman of another major mining group, who toured McLaughlin with me, remarked, 'This is the sprucest mining operation I've ever seen'. Living up to the environmental needs, in fact, has become something of a crusade. Not only is their daily housekeeping impeccable, but the group has long-term plans to reclaim much of the land for grazing, and will build an environmental studies field research station for local colleges at the site. 'We are working at being good neighbours,' said David Fagin, Homestake's president.

While McLaughlin is an object lesson in coming to terms with strict controls, the high costs have acted as a warning light to other mining groups with likely gold prospects in California. They have even prevented some projects going ahead. The Sierra Club, a powerful environmental group that is extremely vigilant in watching mining activities, has also proved a considerable deterrent. Further warning signals came from the experience of Noranda, the Canadian mining group, which faced legal action from the California authorities for alleged environmental damage due to cyanide seepage from the tailings dam at its small

Grey Eagle gold mine in northern California.

Consequently, gold output in California is not surging ahead with the rapidity of Nevada. If California's example is followed elsewhere, then it could slow down the growth of American output. Many mines would just not be developed.

California's controls have not deterred everyone. Several profitable mines have opened. Consolidated Gold Fields' North American arm, Gold Fields Mining Corporation, successfully brought the Mesquite mine in south-eastern California into production in 1986 with an annual output of 4 tonnes. Mesquite is a large shallow open pit, grading only 1.7 grams, but it has a fifteen-year life of proven reserve. Operating costs are comfortably under $200 an ounce, and drilling is expected to confirm satellite orebodies nearby. This corner of California, tucked in close to the Arizona and Mexico borders, has other prospects. Glamis Gold from Vancouver has already been operating a low-cost heap-leaching operation at Picacho for several years that turns in around 700 kilos annually. Farther north, Glamis also expects to bring in the Yellow Astor mine late in 1987.

The terrain of the original California gold rush of 1849 has naturally been thoroughly scrutinised. Sonora Gold Corporation, a Toronto-based company, hopes to have its Jamestown gold mine in Tuolumne county fully operational by 1988 with an output of 3 tonnes annually. Sonora has pinpointed six deposits within its mine lease that together contain well over 100 tonnes of gold and a substantial bonus of silver. However, like McLaughlin, they have a hard time with environment controls, and have to haul much of their concentrate across the California border into Nevada for cyanide leaching. Eventually they hope to avoid this by a new process called thiourea leaching involving bacteria.

Montana: Pioneering Heap-leaching

To find the real pioneers of this new era it is necessary to go to Montana, which has never really hit the headlines as a gold state. Yet the Zortman-Landusky open-pit mines, set into two picturesque neighbouring hillsides in north-central Montana, first showed the viability of large-scale heap-leaching of very low-grade ores.

The project was started in 1979 by Pegasus Gold, a small mining house based in Spokane, Washington. Pegasus had to feel its way. 'There were no experts in heap-leaching gold,' explained Philip Lindstrom, the soft-spoken and courteous mining engineer who is a consultant for Pegasus, as we flew in a small plane from Spokane to the mine. 'We learned by trial and error, and we made a lot of mistakes. Now everyone follows our example. We get over 300 submittals a year to help on other heap-leaching projects.'

Presently, our plane homed in on a small range of pine-clad hills, jutting up from the flat Montana landscape. The pilot circled to check no deer or cattle were straying on the landing strip. Once down, he taxied to a gateway hung with coyote pelts. 'That's to keep the cattle out,' said Lindstrom, as we clambered into a jeep. 'This is real Butch Cassidy country.' Indeed it is. A dirt road led to a cluster of log cabins, a bar with hitching rails for horses out front, and a white-washed jail thoughtfully straight across the street. Nearby a few low prefabricated buildings housed the mine offices, including the assay laboratory which is crucial to the profitability.

Much of the ore at Zortman-Landusky grades less than 1 gram per tonne (compared with 6 grams as the average South African grade, or 75 grams at Japan's new Hishikari mine). Isolating ore from waste is vital. 'We can do 700 gold assays a day,' said Doug Belanger, Pegasus's vice-president for corporate affairs, as we watched women technicians in white lab coats hard at work. 'Then we can plot exactly where the gold is on our computer models of the pits. Grade control is critical to us; we will leach right down to 0.008 ounces per ton (0.3 grams per tonne).' Drilling maps are then tagged in red for gold ore and green for waste, and blasting is carefully plotted to try to separate the two. After blasting, crews stake out the zones with different- coloured flags, so that excavators can load the huge trucks either with ore for the leaching pads or directly for the waste dumps.

Unlike most heap-leach operators, Pegasus does not crush the ore before leaching; the low grade simply does not justify the cost. Instead, shallow depressions have been hollowed out in the hillside below the mine site, lined with pvc sheeting and the jumble of gold-bearing bearing rock is piled up. 'We've got 27 million tons on all the pads,' said Lindstrom, as we climbed above a plateau of ore, the size of several football fields, with hose lines for spraying cyanide solution snaking across. Leaching is slow because the ore has not been crushed. Usually the process takes only a month or two. Here it is judged in years. 'We get about forty-three per cent of the gold out by leaching in the first year, then another seven per cent annually,' said Lindstrom. 'All told we get about sixty-seven per cent over five years.' The actual leaching is intermittent; after the initial thorough dousing, cyanide is sprayed only occasionally. 'We've found it's best to let it sit to mature and ferment, and then when we spray again it washes out a lot of gold.'

This patient process yielded over 3 tonnes in 1986. The best of the reserve has been mined out and output will slowly diminish. 'The glory days are over,' admitted Doug Belanger, 'but we've made good money out of this hill.' Pegasus Gold had also gained good experience, which has given it the confidence to acquire new prospects. They have taken over from Centennial Minerals the Montana Tunnels prospect near Helena Montana, which is an open pit with a cocktail of gold, silver, zinc and lead. Montana Tunnels started up in 1987 and by 1990 is projected to produce nearly 4 tonnes annually. At Florida Canyon in Nevada they have begun another open-pit heap-leach operation scheduled to deliver just under 2 tonnes a year, and are evaluating the viability of re-opening another small open pit at Relief Canyon nearby. Thus by 1990 Pegasus hopes to be producing overall close to 300,000 ounces (9.3 tonnes) annually, which will not only rank them in the top half-dozen US producing groups, but make them the fastest-growing of all in the late 1980s. They are not content to rest either. 'By 1990 we'll have a higher production base,' said Doug Belanger, 'so we can do more acquisitions.'

Another newcomer that is briskly buying up gold mines is American Barrick, the Toronto-based group controlled by entrepreneur Peter Munk. Their advancement is due particularly to a sharp eye for spotting good buys. The real coup that quickly made them a favourite on the mining analyst circuit was the purchase from Getty Minerals of the Mercur mine in Utah. Mercur is set in a cradle 7,000 feet up in the Oquirrh mountains south-west of Salt Lake City. Getty Minerals spent $105 million bringing the mine into production in 1983 only to run into teething trouble getting the right level of recovery from their mill, and then to be taken over by Texaco who decided to sell Getty's non-oil ventures. Barrick snapped up Mercur for a mere $40 million, just when the gold price was in the doldrums in 1985. 'The window was open then and we got Mercur cheap,' admitted Robert Smith, Barrick's chief operating officer. Barrick borrowed 77,000 ounces (2.4 tonnes) of gold, sold it spot to raise $25 million to repay most of the loan taken out to buy the mine, and will repay the loan in gold from Mercur over four and a quarter years at two per cent interest. Meanwhile, the mill has been re-organised, additional carbon-in-pulp recovery circuits installed, and the workers told that Mercur has a secure future as a low-cost producer for the rest of this century. Transformation was swift. Production doubled and costs came down under $200 an ounce. 'The Midas touch', applauded Burnham Lambert, the mining analyst at Drexel. Barrick followed on by securing 26.5 per cent of the efficient Pinson Mine in Nevada, and went drilling at another likely Nevada prospect, Frank W. Lewis, adjoining Battle Mountain's Fortitude mine. They are active, too, at home in Canada, as we shall observe in the following chapter, and caused ripples on the international scene by buying a small stake in Consolidated Gold Fields, and then selling again at a good profit.

This crisp style, bringing in effective management and financial expertise to make the most of gold loans, forward sales or options, is the hallmark of the new gold mining scene in both the United States and Canada. Moreover, houses like American Barrick or Echo Bay Mining operate equally happily on either side of the border. As a new breed, they are less concerned with the original exploration for and proving up of new properties, which can be expensive and abortive. Rather they spot bargains, as Barrick did at Mercur. Echo Bay also picked up the

Sunnyside mine in southern Colorado from Standard Metals Corp, which had filed for bankruptcy. The mine cost Echo Bay $20 million, plus some profit-sharing, which they financed by a gold loan. They have since refurbished the mine and are set to produce 1.5 tonnes a year at $300. Further acquisitions late in 1986 of three small Nevada mines from Tenneco Minerals, a Houston-based conglomerate hard hit by the slump in oil prices, brought Echo Bay's full North America stable up to six mines that will be yielding over 700,000 ounces (21.7 tonnes) by 1990. That will place Echo Bay in the top flight of North American producers, rubbing shoulders with Newmont.

Such manoeuvres are rapidly re-shaping the gold mining industry in the United States. Twenty years ago it was essentially Homestake and by-product producers which made the only running; five years ago there were a plethora of small operators; today nearly seventy-five per cent of output is in the hands of nine mining groups, and by 1990 virtually half could be controlled by just three, Echo Bay, Homestake and Newmont. Although the pack may well be reshuffled again before then.

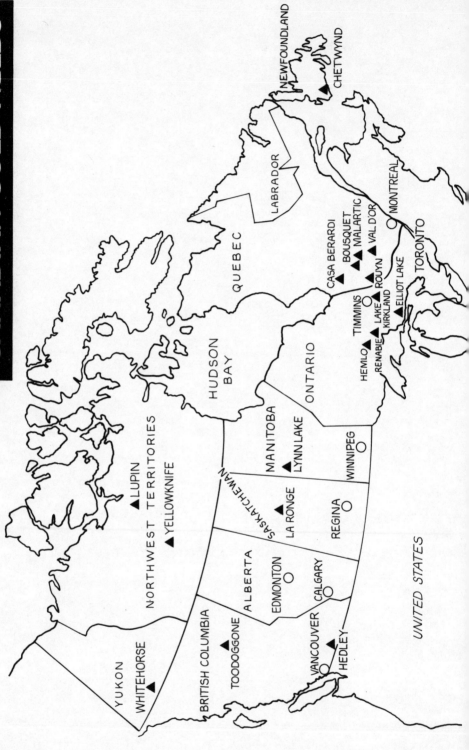

CANADIAN GOLD FIELDS

NEWFOUNDLAND

CHETWYND ▲

LABRADOR

QUEBEC

MONTREAL ○

TORONTO

CASA BERARDI ▲

BOUSQUET ▲
MALARTIC ▲
VAL D'OR ▲
ROUYN ▲
KIRKLAND LAKE ▲
ELLIOT LAKE ▲

TIMMINS ○

HEMLO ▲
RENABIE ▲

HUDSON BAY

ONTARIO

MANITOBA

LYNN LAKE ▲

WINNIPEG ○

NORTHWEST TERRITORIES

LUPIN ▲

YELLOWKNIFE ▲

SASKATCHEWAN

LA RONGE ▲

REGINA ○

ALBERTA

EDMONTON ○

CALGARY ○

YUKON

WHITEHORSE ▲

BRITISH COLUMBIA

TOODOGGONE ▲

VANCOUVER ○

HEDLEY ▲

UNITED STATES

Chapter 4

CANADA: THE HEMLO FACTOR

The Trans-Canada Highway north of Lake Superior is a lonely, beautiful road, mile after mile of spruce and birch trees, of tumbling streams and peaceful lakes, broken only by occasional small communities like White Rock and Thunder Bay. Moose and bear roam the woods. The scenery changes, however, for a mere mile near the town of Marathon as the headframes of three new gold mines appear, jostling for space right by the highway. First, as you drive east, the yellow headgear and mill of the Page-Williams mine, towering over Moose Lake, then the bright blue of Golden Giant and finally, across Cedar Creek, the tan buildings of Teck Corona's David Bell mine. This is the Hemlo gold field, or 'camp' as Canadians usually say, which is the biggest advance in Canadian gold this century. By the time the three mines are at full capacity in 1989, Hemlo will yield close to 1 million ounces (31 tonnes) annually for a life-span of at least twenty years.

In sharp contrast to Australia or the United States, where expansion has come from a host of small open-pit mines, the new dimension in Canada's gold comes primarily from the three underground mines at Hemlo. 'In Canada it's a different game,' said Peter Allen of Lac Minerals, 'it's hard rock and under-ground.' As Canada's output has doubled in the 1980s from a modest 50 tonnes annually to just over 100 tonnes, Hemlo alone has accounted for sixty per cent of the improvement. The rest has come largely from expansion of new mines at existing gold camps, chiefly in Ontario and Quebec, and from one particular enterprise, the Lupin mine up on the Arctic Circle, which ranks as the world's most northerly mine.

Canadian miners, like those everywhere, were initially galvanised by the high price of 1980. For a generation gold mining had struggled to survive and the industry took in few new recruits, especially on exploration teams. Everyone was too busy with base metals. The mood on Bay Street in Toronto, where most of the mining companies make their homes in steel and glass towers, changed only in the early 1980s. A legion of young geologists started looking over old prospects with a fresh eye. 'The industry has been transformed,' admitted Allen. 'Today we have hundreds upon hundreds of applications from geologists. They know the theoretical basis, they are young, they are keen, they will interface with computers and will work long, dedicated hours when you put them in the field.'

Before long the name on everyone's lips was Hemlo. The site had actually been prospected for over a century, but old-timers had looked for conventional quartz veins housing gold. Few cropped up at Hemlo, so the area had been dismissed. What the new generation of geologists found was a disseminated orebody fitting rather like the meat in a sandwich between two different types of rock. One is a sedimentary rock, composed of material found in a lakebed, the other is volcanic. The gold sits in the contact zone between the two. The word 'orebody' is a trifle misleading; rather, the gold-bearing zone is a thin slice only two or three metres thick slanting sideways down into the ground at a steep angle of between 45 and 60 degrees from a few surface outcrops. Just to complicate matters, it weaves and curves as it goes. The grades, however, are good, mostly varying between 5-15 grams and averaging about 9.5 grams per tonne. Moreover, for all its underground twists and turns, almost all the gold is in what miners call one 'horizon'. 'The unique thing is that ninety per cent of the reserves are in a single horizon from the surface down to 1,500 metres along a strike length of two kilometres,' said geologist Doug Conroy at the David Bell mine. 'Not many gold mines are that consistent with one horizon.'

Hemlo, of course, is three mines cheek by jowl, all cutting into that 'horizon'. Therein lie many headaches. Rationally Hemlo ought to be a single gold mine. The competition to prove up the site, however, left little time for rationality and has ended in a welter of lawsuits. Half the problem is the curvaceous nature of the ore horizon weaving its way underground. By mining tradition and law, a deposit pinpointed on the surface of a claim area may be followed to its origins below the surface. But at Hemlo this often means that a surface outcrop plunges down under the claim area of another mining company. Thus underground the ore at one level may be, say, Golden Giant's, while above it is in the province of David Bell. 'You've got to be aware of your neighbours all the time,' said Jacque Pelletier, the mine superintendent at Golden Giant. 'Some of Page-Williams's ore is just twenty feet off our access road. So we've all got to declare our mining methods. We have to have joint pillar agreements, and we can't have a stope next to their stope.' Sometimes one mine trades a section of its ore for a slice of another's. 'It's one orebody and one mine really,' conceded a geologist, 'but it's turned out at three mines with five shafts and made a mess of a nice orebody. The lawyers get rich.'

· The initial credit for Hemlo goes to two experienced prospectors, Don McKinnon and John Larche, backed up by geologist David Bell, who started charting Hemlo in detail in 1980. Soon they were followed by two Vancouver companies, Goliath Gold Mines and Golden Sceptre Resources, controlled by

Richard Hughes and Frank Lang, who took options in 1981 on two key blocks in the heart of the strike zone. As the real potential unfolded, these junior companies went into joint ventures with major mining houses for the actual development. David Bell gave his name to the Teck Corona joint venture; Goliath and Golden Sceptre teamed up with Noranda, the Toronto mining company, in the race to bring the first mine on stream.[1] Noranda won with Golden Giant by a matter of weeks. 'We had to get 1,000 tons [of ore] a day within two years,' remembers Pelletier. 'We made the mine through winters at -40 degrees (C) and in terrible wind conditions. It wasn't easy, but it was very satisfying when we dropped the first bar of gold in April 1985.'

Golden Giant is now fully in its stride. 'We're up to 2,300 tonnes a day and our peak will be 3,000 tonnes,' said Pelletier proudly, 'and it still looks like a 10-gram orebody over the life of the mine. Sometimes it's better than expected, by half a gram or a gram, sometimes it's only 9.5 grams. We ought to get 330,000 ounces (10 tonnes) in 1987.' Costs are low. During the first two years they were scarcely US$110 an ounce.

Although Noranda and their partners produced the first Hemlo gold, Teck Corona's David Bell mine was scarcely a month behind. Lac Minerals, who had been the third contestant since the earliest days, came in six months later with Page-Williams, the biggest of the Hemlo mines. This mine is named after two geologists who prospected the area in the 1940s and first hinted at its rich potential. Lac had been quick to stake out a large claim in 1981 covering the eastern end of the ore horizon, where the gold reserves are larger, but the grade is slightly less (basically Hemlo's grades improve from east to west). Lac got into action by December of 1985 and extracted an encouraging 225,000 ounces (7 tonnes) in the first year, helped by the fact that they could start with a small open pit where the ore horizon touches the surface. The real scope of Page-Williams, however, is in a broad swing of the ore horizon underground. This will be broached over the next few years by two shafts into what are known as the A and B zones.

'We're still on a learning curve,' said mine manager Gerry Gauthier, when I visited the mine. 'So far with the A zone the reserves are up to expectations, but we'll have a better handle on what's down there in the B zone by late 1987.' As that zone is tapped, so output at Page-Williams will rise rapidly. The target is to mine 1.3 million tonnes in 1987 for about 8 tonnes of gold, then stepping up to 1.8 million tonnes for 12.4 tonnes in 1988, before moving to a peak of 2.2 million tonnes with close to 15 tonnes by 1989 or 1990. Much will depend on the consistency of the grades. 'We've been averaging 6 grams,' said Gauthier, 'but at depth it may be better, and we already know that the grades are higher the closer we go to Golden Giant.'

The first two years of Page-Williams have proved there is still a great deal

1 Noranda's other gold operations include Kerr Addison and Chadbourne.

to learn about the true size and scope of the deposit. Exploration teams are probing into a third, C, zone, where initial grades have been only a modest 3.9 grams. But Gauthier admitted, 'The C zone is untapped, we don't know enough about it at depth.' Mining men like to have something like that up their sleeve. He certainly sees a long future. 'I feel confident we'll be here for a while,' he concluded with a grin, 'at least until I retire.'

The real headache at Page-Williams has not been gold grades, but who owns the mine? It was pioneered by Lac Minerals, one of Canada's most ambitious and forward-thinking mining houses. But in the very early days of the scramble for Hemlo, Lac got a look at some work that its rivals, International Corona Resources, had done on claims on what is now Page-Williams. From that stemmed a lawsuit, which ended up with Corona not getting just damages, but being awarded actual control of the Page-Williams mine in 1986. Lac has appealed the judgement, and meanwhile has continued to administer and develop the mine. If the appeal also goes against Lac, then International Corona Resources with its joint partner Teck, which controls the third Hemlo mine, will dominate the field. They will have Page-Williams at the western end, Teck Corona at the eastern, with Noranda's Golden Giant between. Teck Corona's own David Bell mine is the smallest of the three at Hemlo, but being at the richer eastern end has the highest grades, averaging 11 grams. The first gold was poured in May of 1985. When it is fully on stream in 1988, output is forecast at 188,000 ounces (5.8 tonnes), easing back thereafter to about 150,000 ounces by 1990, when slightly lower grades are expected at depth. For the moment the main work at David Bell is on the upper levels, where the ore horizon is shaped like a dog's leg pointing at the surface. Down in the mine the narrow gold-bearing band, sandwiched between its two host rocks, stands out quite clearly. Because the horizon dips so sharply to the north it plunges through each level of stopes at an angle, first visible at the top left and going out at bottom right.

Geologist Pierre Desautels took me underground on his shift to see how he charted the ore. From the bottom of the shaft we scrambled up an endless series of wooden ladders through an ore-pass until we came out into a high, wide stope lit by arc lamps and filled with the roar of drilling machines chattering needle-like into the roof of the stope. We trekked down this tunnel for a while until the roof came down to touch our helmets. Desautels took a can of white aerosol paint out of a small knapsack, directed his miner's headlamp at the roof and walls, and proceeded to paint in the outline of the ore horizon. You could see it, quite distinct from the other rocks, a band varying from a metre to two metres wide in one corner of the roof. Spraying his paint, Desautels marked out the edges in white and occasionally squirted HG or LG to indicate spots of high grade or low grade. At one point, in some excitement, he actually detected some specks of gold, an unusual occurrence in a disseminated ore where the gold is microscopic. Having charted where the ore horizon came through the roof, he then tracked where it went out low on the side walls, measuring everything carefully with a long tape measure and making copious notes. The drillers working behind him can then segregate gold ore from waste. In the heart of the deposit the ore horizon is consistently visible, but at the edge it 'feathers' out, becomes erratic and peters out. 'It's like the squeezed end of a sausage,' said Desautels, spraying vigorously to indicate work need go no farther.

Such legwork down every stope is gradually building up a more accurate picture of the most significant orebody found in North America this century. Known reserves at the three Hemlo mines already amount to 50 million tonnes. 'A continuous horizon of 50 million tonnes of 6-8 grams per tonne is unique,' said Desautels's colleague, Doug Conroy, as they discussed our underground tour. Put another way, Hemlo is known to contain at least 350 tonnes of gold. Since all three mines are looking for a twenty-year life, plenty more is in prospect, not only from the existing operations. A veritable quiltwork of claims from many other mining groups spreads out around these three yellow, blue and tan headframes set against the lonely sky of the north woods.

'Hemlo means that gold is again central to the Canadian mining industry,' Lac's Peter Allen told me when the camp's true potential was first known, 'instead of being just the poor sister.' Hemlo actually accounts for nearly forty per cent of all economically recoverable gold discovered in Canada in recent years. 'Hemlo represents a substantial part of Canada's gold endowment,' says Donald Cranston of Energy, Mines and Resources Canada. Hemlo headlines, however, should not obscure other initiatives.

Lupin: the Ice Road to Gold

On a rocky knoll jutting up from the tundra of Canada's Northwest Territories, just fifty-six miles from the Arctic Circle, is the world's most northerly gold mine, Lupin. It is a remote place on the shores of Contwoyoto Lake, where only a few Arctic hare, fox and caribou passing on migration keep company with 200 miners drilling and blasting nearly 200,000 ounces (6.2 tonnes) of gold out of the permafrost each year. In winter the miners, working 300 metres underground, have to wear parkas because the air pumped down for ventilation is -20 degrees C (above ground it is -40 degrees).

The story of Lupin, owned by Echo Bay Mines, is not just one of overcoming harsh Arctic winters, but of the sheer logistics of building the mine in the first place and then keeping it supplied with fuel, explosives and fresh food. The entire mine establishment was originally built by an air shuttle in an ageing Hercules that made 1,100 round trips to a gravel landing strip. Every single piece of equipment had to be tailor-made to fit into that plane.

Today, the profitability of Lupin is reflected in the smart 727 jet that shuttles between the company's headquarters at Edmonton, Alberta and the mine. The forward half of the plane is loaded with freight, while a fresh crew of miners going in occupies the rear. A huge hot breakfast, that would put most scheduled airlines to shame, is served. As the plane lands, a neat pattern of orange buildings comes into view, set beside a lake still half-covered with ice floes, despite the permanent daylight of high summer. 'Pete, what's the fishing like?' an incoming miner asks, as we disembark. 'Still too much ice,' is the reply.

Although Lupin (named after the wild flower that briefly dots the tundra in summer) is a new mine, the gold deposit was located by accident in the early 1960s. Two geologists prospecting for International Nickel (INCO) near the lake were plagued by blackflies and decided to seek some respite on a little hill nearby to eat their lunch. Being geologists, they took a few samples and found, not nickel, but a gold outcrop. Neither INCO nor anyone else, however, was interested to develop the property with gold at $35. No one paid much attention for nearly twenty years. Then in January 1979 Echo Bay, an offshoot of IU International, took an option on the INCO property. They were casting round for a replacement for a small silver mine in the Arctic at Port Radium, which was almost mined out. Gold was still hovering around $200 and the deposit was not yet economic. But within twelve months the price of gold and silver had taken off. Echo Bay from then sold forward all the remaining silver due from Port Radium. 'We sold 1.4 million ounces at US$35 an ounce,' remembers Paddy Broughton, the vice-president for corporate affairs. 'We netted C$29 million in two days. That was the equity to go into Lupin.'

The mine opened in 1982, launching Echo Bay as one of the most successful newcomers in North American gold mining. Five years on, they have made Lupin, in the words of one Canadian mining analyst, 'a mine that exceeds even its own expectations'.[2] The group has also acquired significant American operations in Nevada and Colorado (see Chapter 3), which should lift them into third place among North American producers, after Newmont and Homestake, by 1990.

Lupin's birth, financed by the forward silver sale, is a classic Echo Bay approach. They are not a traditional mining house that goes exploring for precious metals. They look for proven deposits or existing mines that are going cheap, and then put in strong management. 'The key is financial management,' explained Broughton. 'We offer innovative financing - gold warrants, gold loans, forward sales.'

In the hostile Arctic environment of Lupin, however, Echo Bay has not only got its accounting right, but succeeded in creating a unique working environ-

2 Laima Dingwall, *Canadian Mining Journal*, August 1985

ment, unlike any I have encountered on another mine. It is rare to visit a mine where the sense of belonging to and working with a team is so strong (the contrast with South Africa's strife-torn mines is stark). The miners at Lupin have voted for a schedule of two weeks at the mine, working twelve-hour shifts seven days a week, followed by two weeks off back at home (ferried in and out by the Echo Bay 727). Everyone I interviewed found it transformed their lives. 'It's wonderful,' said a miner who commutes from Texas. 'You bust your arse for two weeks, and then have two weeks completely off.' A nurse in the infirmary, one of a dozen or more women working regularly at Lupin, told me she went home to Newfoundland every fortnight to look after her elderly mother. Twenty-six weeks' work for a full year's pay pleases everyone, and many also run small businesses back at home. Consequently, labour turn-over is almost nil. On the ground, or rather underground, productivity benefits. 'You really get into your work,' said Graham Clark, a twenty-seven-year-old mine captain, who took me underground. 'We can train people and know they'll stay. Each supervisor down here has his own crew of twenty-two. It's a real team business and they all work together as partners for a few years.'

The gold deposit at Lupin is shaped like a giant compressed 'Z', with three gold zones averaging 12 grams per tonne. (The precise geological definition is that the gold mineralisation is strata-bound in a folded and metamorphosed iron formation.) Initially the mine was billed for a six-year life, with reserves proven down to 400 metres. But as the mine has developed, so deeper drilling confirmed the 'Z' formation plunges beyond 1,000 metres, and access through a deeper shaft completed in 1987 assures that Lupin will be in business for the rest of this century.

The mining may also get slightly easier at depth, as the working areas warm up a trifle. In the first levels of the mine, because of the permafrost, the temperature actually gets lower. The rock is -3 degrees C on the surface, -6 degrees at 87 metres, warms up to zero at 490 metres, and by 650 metres is actually plus 1 degree. This eliminates the cooling needed on conventional mines, but brings its own problem. Very cold air pumped down for ventilation often forms fog in the tunnels.

Back on the surface, life is remarkably civilised. The miners have their own rooms, with television (by satellite) and radio. There are a library, sauna and pool tables. Baseball, soccer and fishing offer relaxation in summer, ice hockey and skiing in winter. The only rigid rule is no alcohol. The food, however, is more than ample. I ate two of the largest and best fresh salmon steaks I have ever had in the cafeteria overlooking the lake.

Fresh food comes in on the 727 three times a week, but the real secret of Lupin's financial success is a unique overland delivery system in winter to bring in heavy basic supplies on a 340-mile-long ice road over frozen lakes and rivers. From mid-January to mid-March a highway is ploughed from the town of Yellowknife north to Lupin, along which convoys of trucks make over 800 round trips laden with 4 million gallons of diesel fuel, explosives, steel rods, chemicals and even 1 million pounds of salt.

The journey takes two days, unless the weather closes in. Even without blizzards it is fraught with danger. The pressure of the trucks roaring across ice four metres thick sets up waves running ahead of them under the ice, so that they have to slow down before they reach the shore to prevent landing on a surge of water running up the beach. Occasionally an axle cracks weak ice, but only one truck has fallen right through since the road was initiated in 1984. (The driver jumped clear.) The savings are significant. Echo Bay calculates the ice road cuts freight costs by $3 million annually, equivalent to $16 on every ounce of gold produced. Such reductions are essential because the logistics of operating in the Arctic mean expenses are about thirty per cent higher than on a conventional mine. The ice road enables Echo Bay to shave its costs to $200 an ounce or a shade below.

Five years' experience of extracting gold profitably in the Arctic has encouraged Echo Bay to look for satellite orebodies close to Lupin that could extend the mine's life into the next century. The tundra around is dotted in summer with the tents of exploration teams from other mining companies. While I was there, five squads of rival geologists were hunting for gold within a forty-mile radius of Lupin. The ice road to gold will be carved out for many more winters.

New Life in Old Camps

In tandem with the pioneering at Lupin and Hemlo, many of the older Canadian gold camps are flourishing. None of the newcomers, for instance, can match the Campbell mine at Red Lake in Ontario, which has been going strong since 1949, when it comes to low costs. The mine has an enviable grade of just over 20 grams a tonne, enabling it to produce around 230,000 ounces (7.1 tonnes) annually at scarcely US$110 an ounce, making it one of the cheapest in the world. A $10-million expansion programme will soon lift milling capacity so that output by 1990 will rise to 240,000 ounces.

The Campbell Red Lake group, part of the new mining conglomerate formed in the summer of 1987 by a merger between Dome Mines and Placer Development, is also pressing ahead with the development of an underground mine at Detour Lake near the Ontario/Quebec border. Detour began as an open pit in 1983, but grades did not live up to expectation and costs were $345 an ounce. Campbell and their partners, Amoco Canada Petroleum, persevered by going underground in 1987. This strategy should transform their fortunes. Operating costs will come down to $260, and production rise to 110,000 ounces (3.4 tonnes) for four years, 1988 through 1991, and then fall back slightly. Campbell has also increased its stable by acquiring from Falconbridge the modern Kiena mine in Val d'Or, Quebec, which yields just over 2 tonnes a year at around $210 an ounce and has reserves to carry it comfortably into the next century. Joint exploration with Dome has also turned up a small deposit at Dona Lake in Ontario, scheduled to open in late 1988 with costs around $200 an ounce.

Dome Mines itself has been modernising its original asset, the Dome mine at Timmins, Ontario, which opened in 1909. 'Dome is a survivor,' said Harry Brehaut, the president. 'We're optimistic about existing reserves, and we're using our resources much better.' A new shaft down to 1,500 metres, completed in 1985, has enabled output to increase over thirty per cent, although operating costs are relatively high at $265 an ounce. But there is plenty of life in the mine if the gold price stays strong and reserves still untapped at greater depths can be reached profitably. Dome mine ought to make its centenary.

The group has to watch costs carefully, too, at its fifty-year-old Sigma mine at Val d'Or in Quebec, which yields around 2 tonnes annually at about $270. Indeed, except for the bargain at Campbell, Dome is vulnerable on costs and accordingly sells up to twenty-five per cent of production forward on gold price spikes to lock in profits. 'We've been chasing the market up with 1,000 ounces at a time,' said Brehaut. 'We particularly want to make sure the Detour Lake expansion is paid for.' Looking ahead, Harry Brehaut is keen to acquire some low-cost open-pit mines in the American West. Through Campbell Red Lake, Dome have already taken a stake in Silver State Mining Corporation, which has interests in the Dee and Tonkin Springs heap-leach operations in Nevada. Tonkin Springs's costs are $90 an ounce, a real tonic after expensive Canadian underground mines. 'In the US the stuff is sitting on the surface,' said Brehaut. 'That's why we're focussing on it. It cost us $5 million just to locate the Dona Lake mine here in Ontario.'

Dome's position has been further enhanced by the agreed (and surprise) merger with Placer Development which created the largest gold-producing group in North America, when Placer's US and Australian mines are taken into account.

The lure of Nevada's cheap open pits has not yet seduced Canada's other leading house, Lac Minerals, which remains firmly committed to exploiting and developing the underground hard rock mines it knows best. Lac's star has been dimmed by the decision (still subject to appeal) that it must hand over the Page-Williams mine at Hemlo to International Corona. Yet when I visited Peter Allen, the president, he showed remarkable resilience and talked energetically of his plans for growth. Lac's best bets are the Doyon and Bousquet mines in Quebec. Doyon, started up in 1980, initially became the largest open-pit gold mine in

Canada, but is now in transition to become a wholly underground mine by 1988.[3] Production should rise to 250,000 ounces by 1990, with costs a modest $160-170 an ounce. Doyon may do even better in the '90s. New exploration at depth as the underground mine is excavated has turned up large tonnages with grades between 11-15 grams.

At the nearby Bousquet mine, the prospect looks even brighter. Lac's exploration division found the mine in 1977 and it came on stream in 1979 at around 90,000 ounces a year. So far the mine has been relatively high cost, at close on $300 an ounce, but the future may be transformed by the discovery in 1986 of a massive gold-bearing zone with grades of 7 grams on the east side of the mine. 'What we are turning up at Bousquet in strike length and width reminds me of Hemlo,' Peter Allen told me enthusiastically. The new find could double Bousquet's output by the end of the decade. Lac are equally confident that output from their Macassa mine and reworking at the old Lake Shore mine at Kirkland Lake, Ontario can also double by 1988. A new shaft down to 2,000 metres and a tripling of the milling facility at Macassa will cut costs to a modest $215. Overall, Allen calculates Lac Minerals will be mining 450,000 ounces annually by 1990 (or 850,000 ounces if it wins the appeal to retain Page-Williams at Hemlo), an increase of almost eighty per cent in just five years. Lac also has its sights on several future prospects. The company maintains one of the strongest gold exploration teams in North America. They have been busy throughout Ontario, Quebec, New Brunswick and British Columbia. Already encouraging results have shown up at Barraute in Quebec, which could open in the early 1990s.

Lac, Dome, Campbell Red Lake and other older Canadian mines like Agnico-Eagle, Aiguebelle Resources, Dickenson Mines and Giant Yellowknife have had to match brisk competition in recent years from newcomers, who know less about exploration and mining but more about finance. American Barrick, the Royex/International Corona Resources alliance, and Teck Corporation, have been pouncing like eagles on bright new prospects, or on poorly managed, high-cost older mines, especially at tough moments when the gold price has been in the doldrums. Teck Corporation from Vancouver, for example, took on INCO's Golden Knight property at Casa Berardi in Quebec, an area reckoned to have excellent prospects. 'Teck is very aggressive and just moves in when they see something they like,' observed David Duval, western editor of *The Northern Miner*. A brash batch of 'junior' mining companies, as Canadians put it, has also staked claims in the old established camps around Timmins and Kirkland Lake in Ontario, and across the Quebec border in Val d'Or, Malartic and Casa Berardi. Canamax Resources Inc. and Consolidated CSA Minerals, for instance, were getting their first full year's production from their Bell Creek joint venture near Timmins in 1987. Eastmaque Gold Mines, a small Vancouver-based company, has

---

3 Lac Minerals operates the mine, and has a fifty per cent share. The partner is Cambior Inc., a company created by the privatisation of the Quebec government's mining interests.

started a tailings operation at Kirkland Lake, where American Barrick is busy developing its Holt-McDermott property which is due on stream in 1989 at 100,000 ounces (3.1 tonnes) in 1989.

Holt-McDermott will be the first gold mine actually initiated in Canada by Barrick, the Toronto-based company controlled by entrepreneur Peter Munk, which has made a rapid shift from the oil and natural gas business to precious metals. Barrick first got into Canadian gold by acquiring the old Camflo mine at Malartic in Quebec. 'Camflo gave us the technical background,' Robert Smith, Barrick's chief operating officer, told me. 'We've blended that with our own innovative financing to try to get the best of both worlds.' Camflo's purchase showed that financial know-how at work. Barrick sold forward $40.1 million of the mine's future output to Gold Company of America, an alliance between itself and Prudential-Bache Securities, to raise most of the purchase price. The added asset for Barrick is that Camflo controls several exploration claims in the surrounding Malartic and Val d'Or camps, thus securing the group a good foothold in one of Quebec's best gold locations.

Barrick also have a fifty per cent holding in the Renabie mine in northern Ontario. The mine opened originally in 1947, but has had a chequered history; it closed in 1970, reopened briefly in 1975, started again in 1982 after modernisation but was ailing with high costs. Barrick then stepped in on a joint venture rescue with Royex Gold Mining Corporation. The partners have pared down operating costs from a disastrous $374 an ounce in 1984 to around $240. Deep drilling has confirmed new reserves, so that Renabie, which yielded a mere 16,000 ounces in 1984, will produce a steady 46,000 ounces into the 1990s. As at Camflo, Barrick came up with original financing by raising $15 million through an offering of units in the Renabie Gold Trust, to which a share of output is assigned each year. The Trust's attraction is its share rises in line with the gold price. At $300 a mere three per cent of production is dispersed to unit holders, but at $400 it is four per cent and so on to a maximum of ten per cent if gold is at $1,000.

Royex Gold Mining Corporation, Barrick's partner at Renabie, is also emerging as a new force in North American gold, combining such Nevada open pits as Pinson and Preble with a strong presence in Canadian underground mines. Royex is allied to International Corona Resources and, since a complex restructuring of both in 1987, is now the parent company. Peter Steen, a crisp and efficient mining man of long experience, is chairman, president and chief executive officer of Royex and Corona. Their strongest cards are, of course, the David Bell and Page-Williams mines at Hemlo (assuming the judgement in their favour against Lac is upheld). Steen hopes to build on that solid base. 'We're looking long term to be one of *the* low-cost producers in North America,' he told me. 'We don't have to rush around the world, we have sufficient low-cost reserves here for a long time.' He is convinced that plenty more will be found. Royex has particular hopes for exploration projects at Casa Berardi in Quebec, but is branching out across Canada as the horizon for gold broadens from the customary areas of Ontario and Quebec. They have a slice of the Jolu project in northern Saskatchewan, which is showing spectacular initial grades of over 17 grams per tonne. Royex also has close links with Mascot Gold Mines, which manages two old gold mines in British Columbia which are being rehabilitated. The most important is the Nickel Plate mine at Hedley, 150 miles east of Vancouver, which was re-started as an open pit in 1987 and is scheduled to produce 160,000 ounces (5 tonnes) in 1988, and has a confirmed life at least to 1998. Although Nickel Plate is relatively low-grade at just under 5 grams per tonne, operating costs are forecast to be only $119 in 1988, then rising towards $180 as the mine gradually shifts underground.

British Columbia, incidentally, is enjoying a gold revival. Blackdome Mining Corporation, a small Vancouver-based company, has opened the Blackdome mine near Clinton, that will produce about 40,000 ounces annually. The Toodoggone area in northern British Columbia is bristling with new claims.

Keeping pace with Canadian expansion is becoming almost as hectic as in the United States. The new faces of the late 1980s include BP Canada's Hope Brook Gold Inc., launching Newfoundland's first gold mine at Chetwynd; Dumagami Mines opening at Bousquet in Quebec; Terra Mines at Bullmoose Lake near Yellowknife in the North-West Territories; and Emerald Lake's Golden Rose at Sturgeon Falls, Ontario. In the province of Manitoba, Sherritt Gordon is opening MacLellan, and Granges Exploration has Tartan Lake; while just over the border in Saskatchewan, a clutch of placer deposits has been pinpointed along the Churchill river, and the Saskatchewan Mining Development Corporation is bringing in Star Lake. Up in the Yukon, Total Erickson Resources is aiming for 77,000 ounces annually at its Skukum mine near Whitehorse. This spate of new projects alone could add another 500,000 ounces (15.5 tonnes) to Canada's output by the end of 1988.

Vancouver – the Financial Seed Bed

The springboard, particularly for many of these smaller mining companies, has been the Vancouver Stock Exchange, which has emerged as a major source of venture capital. The lure of Vancouver quite simply was that its rules, at least until 1985, were more liberal than most other North American exchanges. This enabled all kinds of marginal mining ventures to be floated and talked up a storm, often on the basis of a couple of lucky drilling cores. Many amounted to nothing, but Vancouver became the essential seed bed for many viable projects. Much of the initial money for Hemlo was raised there.

'I'm a creature of the Vancouver Stock Exchange,' the chairman of one small mining company said, looking out over the broad sweep of Vancouver's shoreline from his skyscraper office. 'It is a source of madcap financing. The rules were liberal so this brash exchange attracted all of the best and the worst, but enough discoveries have been made to keep the ball rolling.' Half the art in understanding Vancouver is to know who is teamed up with who, and what promotion they might be doing on each other's behalf. 'You have a network of junior companies who crawl into bed with each other,' a Vancouver mining editor explained. 'And then if something is found, they've got a finger in it.' A mining executive added, 'It's a buyer-beware market, but the investor going in there knows he may lose every penny. It has to be a free-wheeling market to be successful.' Punters have been stumping up anything between $200 and $300 million annually, pushing the turn-over on the exchange to over 3 billion shares in the peak years. Vancouver has become such a magnet for gold share players world-wide that it has occasionally run second to the New York Stock Exchange for the biggest volume in North America. 'We get a lot of European money, especially from France and Germany, and a lot from Hong Kong,' an exchange member told me, 'and seventy per cent of the money raised is now leaving Canada to explore for gold in the US, Australia or Brazil.'

Within Canada itself, gold investment has also been encouraged by a flow-through share-financing scheme initiated by the government in 1983 to stimulate mining. This enables an investor, private or corporate, to write off investment in mineral exploration within Canada against other income. 'All kinds of wealthy doctors and dentists subscribe,' a mining executive explained, 'and they get 90 cents back on every $ they put in.' In the space of three years, almost US$550 million was raised in flow-through shares, of which three-quarters was spent on gold. 'Flow-through shares have made all the difference,' American Barrick's Robert Smith told me. 'Without them there would have been much less exploration in gold.' Noranda's Paul Warrington added, 'It's kept the exploration going. Every diamond drill available is now working in the province of Quebec on lots of orebodies we've known about for years. With that kind of money pouring in you must hit something.'

The outlook for Canadian gold mining, therefore, remains good. Output has already doubled, new mines or extension of old ones are still coming on stream, and the diamond drills are probing for more. Canadian gold production shows no sign of peaking, and can move onward towards 120-130 tonnes during the 1990s. Moreover, unlike Australia or the United States, where many small open pits have only a three- to five-year life, the Canadians are finding serious underground mines whose life-span is judged in decades. As Peter Steen of Royex summed it up wistfully, 'There's lots of orebodies like Hemlo sleeping out there.'

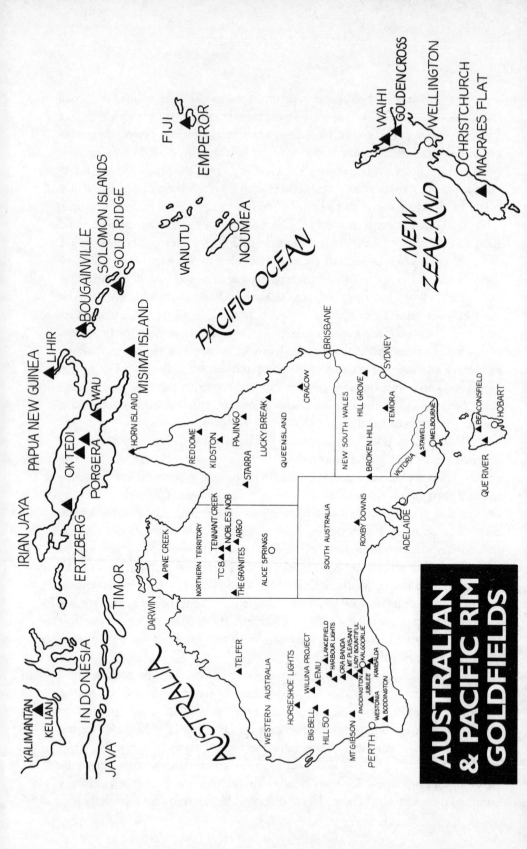

AUSTRALIAN
& PACIFIC RIM
GOLDFIELDS

Chapter 5

AUSTRALIA:
NEW LIFE ON THE GOLDEN MILE

'Man, it's out of control, and there's tons of money being made,' said John Jones, the chairman of Jones Mining and director of the Australian Gold Mining Industry Council. We were sitting in his Kalgoorlie office looking out on the main street, which is named after Paddy Hannan, who first found gold here in Western Australia in 1893. Kalgoorlie has changed remarkably little since its first gold rush almost a century ago. Across the way were the low buildings of Montgomery Bros, General Drapers to the Goldfields since 1893, and the freshly painted ornate wrought-iron balconies of the Old Australia Hotel offering '1st Class Accommodation'. New signs had sprouted, too, over hastily refurbished shop fronts - Mount Martin Gold Mines, Central Kalgoorlie Gold Mines N.L. and Great Western Gold Exchange. Hannan Street leads directly to the reddish tailings dumps of the Golden Mile, which press in on the edge of town. Over 40 million ounces (1,244 tonnes) have been extracted from the Golden Mile in just under a century. Today it is as active as ever. Indeed, the whole of this El Dorado, once a jumble of individual underground mines, is being gouged out in a series of large open pits that are gradually sifting through every square inch of ground in an area scarcely two miles long and half a mile wide. (Correctly, it is the Golden Square Mile.)

This mining boom was undoubtedly triggered by the high gold price of 1980. It has been driven by new technology, particularly carbon-in-pulp treatment which unlocks clay-rich oxide ores that could not be mined economically by the old-timers. Kalgoorlie is the heart of the re-birth of Australian gold mining. The Eastern Goldfields of Western Australia, centred around the town and spreading north and south in a ribbon of new mines over 300 miles long, will produce close to 70 tonnes of gold in 1987, compared to just 17 tonnes in all Australia in 1980. The whole of Australia is forecast to yield comfortably over 100 tonnes in 1987, according to Pat Gilroy of the Gold Producers' Association in Perth. With no less than 250 gold mines now open, in preparation or listed as promising prospects, it could climb over 130 tonnes in the next year or two.

The rapid growth is possible because most newcomers are operating open pits that can be brought into production in a matter of months once drilling confirm's an orebody. The record is held by Mount Percy, just outside Kalgoorlie, which came on stream in a mere twenty-seven weeks. 'It's developing very, very fast,' said Graham McGarry of the Amalgamated Syndicate. 'The next twelve months will stun everyone, and 1988 could be the most significant year yet.'

The boom has an added incentive because gold mining profits are not taxed in Australia. A threat to impose tax in December 1986 by Australia's finance minister, Paul Keating, was beaten off by a determined mining lobby. They argued successfully that gold mines throughout Australia's history (and also in Canada and the United States) have fulfilled a unique role in opening up remote frontier regions, thus creating jobs and stimulating regional economies.

That argument certainly holds for Kalgoorlie. A decade ago it was a symbol of the despair of Australia's gold industry. Not a single mine on the Golden Mile itself was open, and only one nearby mine, Mount Charlotte, was working, picking at a few scraps of high-grade ore. (The Golden Mile in its heyday often yielded 30-60 grams per tonne, and occasionally 300-500 grams.) A small nucleus of miners remained in town, fearing their redundancy notices any day. Salvation came with the gold price and a fresh scramble to assemble leases on the Golden Mile, which had once been divided among fifty companies. By early 1987, most of the Mile was effectively in the hands of just two key groups, with Kalgoorlie Mining Associates (KMA), an alliance of Western Mining Corporation and America's Homestake Mining, rivalling North Kalgurli Mines, controlled by Perth entrepreneur Alan Bond.

Bond, a relative newcomer to gold, is intent on building a mining empire to match his others in beer (Swan Lager), newspapers and television. 'Bondy has definitely got the gold bug,' said a mining analyst. 'Alan Bond wants a 400,000-ounce-a-year operation,' admitted Colin Agnew, chief executive of Dallhold Resources Management, which runs Bond's gold ventures. Already, besides North Kalgurli, he has a stake in the Mount Percy mine just north and the Jubilee mine just south of Kalgoorlie. The heart of his strategy, however, is to open up what he calls 'The Big Pit' to engulf the whole northern section of the Golden Mile itself. The Big Pit would be a mile long, half a mile wide and over 600 feet deep, from which ore could be mined at a rate close to 1 million tonnes annually. That might be just the first phase. The logical extension would be a joint venture with Kalgoorlie Mining Associates next door for an even bigger pit to effectively spoon out the entire Golden Mile to a depth of nearly 1,000 feet. 'It is still nebulous, we're just starting to think about the problems,' admitted Geoff Hopkins, Western Mining's chief geologist in charge of its KMA interests, 'and it would take ten years to flower.'

The project may never mature, but the concept alone underlines the dimension of change here and elsewhere in Australia. 'North Kalgurli and KMA are now extracting close to 400,000 ounces a year from the Golden Mile,' said Doug Daws of Kalgoorlie Gold Services, which is building a new refinery to handle the gold. He expects plenty of custom from other new projects throughout the neighbourhood. Daws rattled them off: 'Jubilee, New Celebration, Hannan South, Mount Martin, Mount Charlotte, Mount Percy, Mount Pleasant, Paddington, Black Lady, Blue Funnel, Lady Bountiful, Grant's Patch, Ora Banda, Broad Arrow, Bardoe, Kanowna - and I'm talking producing mines only.' Another clutch of new properties around Kambalda, fifty miles to the south, are also prospering.

Western Mining began mining nickel at Kambalda in 1966 and then found an abundance of gold around Lake Lefroy. They are already tunnelling under the lake from an island and have a scatter of open pits around the shore. Mining gossip in Kalgoorlie, where everyone meets after work in such raucous pubs as the Boulder Block and the Main Reef, suggests Western Mining may be planning the world's largest open pit, nearly seven miles long, to open up a mammoth orebody running south-east from Lake Lefroy. Western Mining is cautious on its plans, but does admit they will get at least 200,000 ounces (over 6 tonnes) annually from Kambalda region.

The spate of exploration throughout these Eastern Goldfields around Kalgoorlie and Kambalda becomes apparent in the surrounding bush country. Doug Daws took me north out of town through mile after mile of scrub and eucalyptus trees heading for the next community, Leonara, 150 miles away. 'Look at all those drill holes marching off to the right and left,' he said, 'every square inch of this terrain is prospective ore, and eighty per cent of it is covered with overburden and only the twenty per cent of exposed rock has been properly examined by old-timers.' We passed literally thousands of boreholes marked with coloured tape, with little yellow plastic bags of surplus sample stacked nearby. Occasionally we noted a deep slit trench which geologists had dug hoping to open a 'window' on a suspected orebody below.

Presently we came upon a big water storage tank labelled 'Paddington', that signalled Pancontinental's new mine which opened in 1985 and is already the seventh largest in Australia, yielding 2.8 tonnes annually. Unlike many open pits around Kalgoorlie which have only a three to five-year life-span, Paddington's huge orebody will last easily to the year 2000 and current costs are only US$170 per ounce. Just beyond Paddington, we turned off the paved road and bounced along deserted dirt roads, first to a new open pit named Lady Bountiful. The ore here is so soft it can be excavated without explosives. The floor of the pit was an object lesson in new grade-control techniques. Samples had been taken from traverses, assayed and then fed into a computer which prints out a clear floor plan of the pit, marking the gold grades, the marginal and the waste. This plan was then translated into neat white lines of lime along the pit floor. All the excavation crews had to do was dig out separately the gold ore and the waste.

We drove on through more bush country, populated only by kangaroos, to another pit, Mount Pleasant. 'This has been christened the Golden Kilometre,' said Doug Daws, 'and it is important because not only has it been located beneath considerable overburden, but it is a new type of orebody running from east to west, compared to almost all other mines in the Eastern Goldfields where the orebody runs north-south following the Boulder Fault for about 300 miles.' Mount Pleasant (owned by a consortium of Elders Resources, Square Gold, Southern Resources and Geometals) excites geologists because it breaks the rule. It has a core of 4.8 grams per tonne surrounded by a low-grade halo, and raises an important question: is it a maverick or are there more such east-west trends around?

The Geological Detectives

That is the key issue for the long term. What completely new, unsuspected orebodies are likely to be located around Kalgoorlie? So far most new mines have been on orebodies actually located by old-time prospectors, but uneconomic with their limited technology. The long-term way ahead has to be completely fresh finds. Without them, the current surge around Kalgoorlie will peak by 1990, and output will fall thereafter as open pits are worked out.

So what have the geological detectives in mind? My best lesson in the geology of Western Australia came from Geoff Hopkins of Western Mining. He spread out a geological map of Western Australia on a conference room table in his Kalgoorlie office. The different rock strata were marked in a rainbow of colours. Four long, snaking strips of green stood out clearly, running generally north-south through a flat, weathered Pre-Cambrian shield. 'There are four key belts of greenstone that are host for the gold,' Hopkins explained. 'The first is from Norseman 300 miles north to Wiluna, with Kalgoorlie half way; the second is the Southern Cross Belt with mines like Marvel Lock and Westonia; then there's the Murchison Belt from Hill 50 through Big Bell to Horseshoe Lights; and finally Boddington south-east of Perth.' Big Bell, run by Australian Consolidated Minerals and Placer Pacific, and Boddington, owned by Reynolds, Broken Hill Proprietary, Shell and Kobe, are two of the biggest prospects in Australia. Boddington, starting in 1987, should produce over 150,000 ounces (5 tonnes) annually. Big Bell, due in 1988, is forecast at 240,000 ounces (7.5 tonnes) and will become Australia's biggest mine.

The significance of these mines is that both are large low-grade deposits: Big Bell is 2.7 grams per tonne and Boddington a mere 1.7 grams. They open up new horizons, as Geoff Hopkins explained: 'In the past, the classic bonanza gold camp was something like Norseman, with at least 10 grams per tonne. But now technology is bringing home with a thunderclap the possibility of cut-off grades down to 1.5-2 grams, which no one had looked for. Suddenly what you'd call an orebody has changed and our exploration targets have switched. The concept of an open-pit, carbon-in-pulp, treated orebody that is payable is radically altered.' Are there large deposits like this in the Eastern Goldfields? Hopkins believes one signpost is Mount Gibson, a modest orebody grading 1.9 grams a tonne, which started late in 1986. 'It's a laterite orebody [i.e. against the main north-south flow],' he explained, 'that would have been untouchable a few years ago.'

A sustained high level of production in Western Australia into the 1990s will depend very much on finding more Mount Gibsons, Big Bells or Boddingtons. Certainly there is no let-up yet in the exploration. Encouragement comes from the fact that it is not just big mining groups that are showing good profits. On the contrary, many small groups have chalked up remarkable successes. Hawk Investments, for instance, run by two lawyers Chris and Peter Lalor, has revived the old Sons of Gwalia mine at Leonara (where Herbert Hoover, later President of the United States, was the first manager in 1898). The Lalor brothers started out treating the old tailings dumps in 1983, then drilled the orebody, finding payable grades close to 4 grams per tonne. The mine now yields nearly 2 tonnes a year and the Lalors have watched shares in their mine soar from A$0.50 to almost A$8.00.

The essence of the business, however, is a very fast return on capital. The current Australian boom contrasts with the long-term investment involved in developing a new South African mine or Canada's Hemlo deposit. 'Gold is the only shining light on the mining horizon, but it's a five-year route only,' Denis Horgan, chairman of Barrack Mines told me in Perth. 'If we can get a return in five years, then we go.' Barrack set out with Horseshoe Lights in the Murchison Field of Western Australia, brought in Wiluna Hard Rock nearby in early 1987, and is continuing farther afield with the Croydon mine in Queensland due in late 1987.

Queensland Turns Up Trumps

Barrack's Queensland mine foreshadows the next major step forward in Australian gold mining. So far Western Australia has made the running, and contributed seventy per cent of the gold. Going into the 1990s, most miners agree Queensland has the best potential.

Northern Queensland just touches the 'rim of fire' which girdles the South Pacific with a string of rich volcanic or epithermal gold deposits (see Chapter 6). Already the Kidston mine, operated by Placer Pacific, has paved the way, yielding close to 7 tonnes annually since commissioning in 1985. Kidston instantly became Australia's largest mine (at least till Big Bell starts in 1988) and actually paid back its capital costs in the first year. Kidston is a large low-grade deposit at 2 grams per tonne, with a life-span through most of the 1990s and very low cash operating costs of around US$120 an ounce (making it one of the cheapest in Australia).

The tempo in Queensland is increasing. Elders Resources' Red Dome mine came fully on stream in 1986 at nearly 2 tonnes annually. Two significant mines open in 1987. Pajingo, owned by Battle Mountain Gold from Houston, Texas on their first overseas foray, is a high-grade gold-silver deposit yielding over 10 grams a tonne of gold with a bonus of 40 grams silver. The mine starts at 2 tonnes a year but could expand. The second mine, Cracow, in which Costain is the largest shareholder, is lower grade but promises 1.5 tonnes a year. At least twenty-four other prospects in Queensland were being developed or showed promise in mid-1987. 'Queensland will produce about 600,000 ounces this year [1987],' Pat Gilroy of the Gold Producers' Association predicted. 'That's fifteen per cent of Australian output.' These are early days. Ross Louthean, editor of *Gold Gazette*, which chronicles the gold boom, notes, 'Few doubt that Queensland is going to produce major deposits; the bright new face this year was the Strategic-Shell joint venture at Woolgar, where a number of large epithermal gold zones have been outlined at surface.'[1]

A challenge to Queensland also comes from the Northern Territories. In the Pine Creek area, just south of Darwin, Renison Goldfields already has a new mine operating at 2 tonnes a year and seven other likely orebodies in the vicinity are being tested. The best bet may be BHP and Noranda's large low-grade deposit at Coronation Hill (which also contains twenty per cent platinum ore). Meanwhile, in the lonely heart of Australia, midway between Darwin and Alice Springs, Peko-Wallsend's Tennant Creek is well established and Norseman Gold Mines' TCB came on stream in 1987. TCB is a small but exceptionally high-grade deposit clocking up 50 grams per tonne, that will yield 75,000 ounces (2.5 tonnes) annually for a few years. All told, the Northern Territories could soon produce 12-15 tonnes a year.

By comparison to Western Australia, Queensland and the Northern Territories, the gold boom in other states is off to a slow start. The lack of gold mining tradition (at least for over a hundred years), and tougher environmental controls in the more populated states of Victoria and New South Wales, have contained exploration. The picture is changing slowly. In New South Wales, where the first Australian gold boom began in 1851, seven projects opened in 1986, including three treating tailings from the original gold boom. Of the actual new mines, Temora, owned by Paragon Resources, is the largest. It is a medium-sized pit at 2.5 grams per tonne, yielding around 1 tonne annually. 'It's a very, very disseminated orebody,' Stan Lewis, Paragon's managing director, told me, 'but we

---

1 Ross Louthean reporting in *Australian Business*, 3 December 1986, p.57

are looking at a ten-year life, and we are confident there's lots of gold there. We'll get our initial capital back in twelve months.'

Equally there must be lots of gold in Victoria, scene of the 1852 gold rush around Ballarat and Bendigo. Historically Victoria has produced almost as much gold as Western Australia, but today the miners are less welcome. 'The "Greens" live here,' said a senior mining executive in Melbourne, 'so Victoria is not as conducive to developing gold mines. There is plenty of gold around Bendigo, which was a major field before 1918, but we have to drill with silencers and only at specified times.' Consequently only one modern mine, Stawell (owned by Central Norseman), has been opened up, producing one tonne. Modest beginnings have also been made in South Australia, where Western Minings and BHP have just opened Roxby Downs, and Tasmania where Beaconsfield and Quo River are on stream.

Made-to-measure Financing

The sheer speed with which the Australian gold boom has gained momentum owes much to innovative financing. Three local groups, Mase-Westpac (the bullion arm of Westpac Bank), MacQuarie Bank and Rothschilds Australia have been the pioneers.

They have initiated gold loans to mining groups against forward sales of part of the eventual production of a new project. The borrowed gold is sold, giving immediate cash flow to the mining group to develop the project more cheaply than borrowing money. Gold loans are typically available for three per cent. So a mine borrows, say, 1 tonne of gold and sells it spot, while committing part of its first year's or eighteen months' production forward to repay. (Usually the interest is in gold too.) Gold loans caught on so fast that at least half of all new mines have been underwritten in this way. In one of the first big deals, Mase-Westpac set up a A$40 million gold loan for Pancontinental's big Paddington mine in Western Australia.

The popularity of loans even surprised the bankers. 'We went from a standing start to 300,000 ounces loan in eighteen months,' recalled Hank Tuten, Rothschilds' managing director in Sydney. Soon the banks were tailoring comprehensive packages, offering not only loans but purchase of all production once the mine came on stream, and the organising of refining and marketing. Rothschilds, with their merchant banking experience, knitted together complete 'off-the-shelf' mine packages for smaller operators in conjunction with the construction engineering group Minproc (Mining and Engineering Service Pty). 'We were trading gold, so we decided to go out and find clients and see what they needed,' said Tuten. 'With Minproc we helped with feasibility studies, built the plants and then picked up the physical gold. We've provided fourteen turnkey plants.'

This nursing by experienced bullion banks has helped many small companies. 'We did Temora with a gold loan,' Stan Lewis of Paragon told me. 'The interest was only three per cent, so it was cheap borrowing. We've sold 1,500 ounces a month forward to December 1988 to repay.'

The catch, of course, was that if the gold price rose the bullion bank took a nice profit when that forward commitment was fulfilled, while the unfortunate miner was locked into a fixed price. To restore confidence, the banks now offer a guaranteed floor price, with a share of the additional profit if the price is above it. 'The floor may be guaranteed at A$600,' explained a banker. 'Between $600 and $630 the miner will get the real price, and at $630 or over the mine gives thirty per cent of the extra to us.'

The gold loan, 'that locomotive of Australia's gold mining resurgence' as one mining writer put it,[2] has meant that Australia's new output had some advance impact on the market, because of accelerated sales from potential mines that did not yet exist. In 1986, for instance, when Australia's real output was 75 tonnes, close to 100 tonnes of gold came to the market because of loans.

The vogue for gold loans and the accompanying all-purpose packages have lured other banks to Australia from the United States and Switzerland, and the idea is catching on in North America. In the duller trading market of the mid-1980s, bullion banks have welcomed the chance to underwrite a booming mining sector.

The banks also foresee plenty of action as the Australian mining industry is restructured into larger groups, with greater appeal to international gold share investors. The emerging contest is particularly between the more traditional mining houses like Broken Hill Proprietary (BHP), CRA and Western Mining, and entrepreneurs like Alan Bond, Kevin Parry, Laurie O'Connell and Bruce Judge, who have made their fortunes in industry or retailing, but now see gold as an irresistible magnet. 'In Australia it's been entertainment, beer, the media and now gold,' said Don Morley, finance director of Western Mining in Melbourne. 'That's where the money is.'

Establishing Gold Corp. Australia

How long will the profits last? Certainly there is still plenty of momentum for a year or two. At the Perth Mint I saw a graph of the assaying they have undertaken for prospective mines. In the second half of 1986 it shot up from 7,000 assays a quarter to over 12,000. The Mint itself, long a rather sleepy institution (a visitor some years ago was offered a Foster's lager by the director, who took it out of his safe), is being totally modernised to cope with the spate of production. 'We have official capacity for about 50 tonnes annually, but we're now pushing through at

2 Bruce Jacques, *Australian Business*, 3 December 1986, p.58

the equivalent of 90 tonnes for the last two months,' Don Mackay-Coghill, the new chief executive of Gold Corp. Australia, told me. Gold Corp. has been established by the Western Australia Development Corporation to revitalise the refining and marketing of the state's gold. Mackay-Coghill, previously chief executive of the International Gold Corporation in South Africa, is planning two new refineries: one at Perth Airport with 130-150 tonnes a year capacity, the other in Kalgoorlie to handle 30 tonnes annually. They will produce 9999 gold, including kilo bars, so that Australia can become for the first time a significant supplier in the regional markets of South-East Asia. Prices will be competitive: gold *loco* Perth is usually 30-70 cents an ounce cheaper than world markets. Gold Corp., besides becoming a gold trader, is also behind the Nugget, Australia's bullion coin challenge to Canada's Maple Leaf and America's Eagle.

The investments being made in new mines, refineries and in marketing strategies clearly show that Australia has come back in the major gold league and intends to stay there for some while. At least through to 1990, production will exceed 100 tonnes annually and could even peak nearer 130 tonnes. If Queensland lives up to expectations it might just go higher.

Gold mining, however, is a fickle business. Two possible clouds on the Australian horizon could be more pressure to introduce a gold mining tax, and the need for mining companies to decide if it is worth going underground when the present open pits are depleted or cannot safely be worked at depth. Everyone breathed with relief when the tax threat was abandoned in December 1986. But it could resurface. 'Good things don't last for ever,' the finance director of one leading mining house admitted. 'No party will dare say anything for a year or two, but by about 1989 they will tax it - they can't help it.' Such a tax could cool enthusiasm very fast, especially for marginal projects. The decision to go underground in some mines might also be cancelled. The boom must end one day: Australian output cannot accelerate year after year. By the early 1990s the peak will have passed, but it will still leave Australia on a high plateau of production never conceived of a few years ago. For the rest of this century, output should stay over close to 100 tonnes annually and perhaps a little more. That is five times the level of the 1970s and early 1980s, yet another measure of the gold mine revolution.

Chapter 6

PACIFIC BASIN:
THE RIM OF FIRE

The greatest challenge for gold miners for the rest of this century is a broad arc of volcanic rocks swinging round the Pacific Basin from Chile through Fiji, New Zealand, the Solomon Islands, Papua New Guinea to Indonesia, and then darting up through the Philippines to eastern China and Japan. The crescent is called 'the rim of fire', and contains scores of epithermal gold deposits. 'Epithermal' means 'close to heat' and these orebodies are found on the surface near to ancient vents or volcanoes. The gold has become concentrated as it passes through the heated rocks of a volcanic system, then, when it cools on the surface, it often forms very rich deposits - almost a golden cap sitting atop a volcano (although this may often have weathered over time and the gold spread over the surrounding area).

The potential from 'the rim of fire' has been known for some while. Not only have huge porphyry copper deposits at such places as Bougainville and Ok Tedi in Papua New Guinea yielded handsome bonuses in by-product gold, but the Emperor mine on Fiji, a typical epithermal deposit, has been operated since 1932. In the Philippines, Benguet, Atlas and Philex have been mining porphyry copper-gold deposits, and Benguet and Lepanto have been tapping epithermals. More recently, two new epithermal orebodies, El Indio in Chile and Hishikari in southern Japan, have revealed spectacular grades. Hishikari, operated by Sumitomo Metal Mining, has some sections at 150 grams per tonne and has been mined since 1985 at an average grade of 75 grams, with a twenty-year life at 6-7 tonnes. Good grades are one attraction, but other epithermals promise huge low-grade deposits. At the Porgera prospect in Papua New Guinea, which is being developed by Placer Pacific, reserves of nearly 80 million tonnes, grading at 3.7 grams, have been identified. When Porgera comes on stream in the early 1990s it will produce nearly 25 tonnes a year, making it the largest mine outside South Africa. Placer has another deposit almost as large, but with lower grades, in its sights at Misima Island, just south-east of the PNG mainland, which could be in production by 1989. Meanwhile Kennecott, in partnership with Niugini Mining, is busy proving up a mammoth deposit of close to 140 million tonnes on another island called Lihir. Although much of Lihir is low grade (at 2.6 grams), about one-third of the orebody is 4.5 grams. So the mine is billed to produce perhaps 12 tonnes from 1990. All told, Papua New Guinea has five world-class discoveries already.

Coming to grips with the rim of fire is no picnic. Most deposits are in tropical rainforests, in mountainous country or on remote islands with no infrastructure. 'Epithermal means hot, wet, mosquitoes, malaria,' said a jaded geologist looking back on a lifetime of scouting Papua New Guinea, Indonesia and the Philippines. 'The physical costs in people and equipment are enormous. Try to get a helicopter or a drilling contractor - you can't find one.' Just to make it tougher, the metallurgy of unlocking the gold from sulphides and silica in epithermals like Porgera and Lihir calls for such expensive techniques as pressure-leaching or roasting, which raise a host of environmental hazards. 'With a roaster you have a real problem of acid rain,' admitted a mining engineer, 'which in a wet tropical climate like Papua New Guinea would devastate the sweet potatoes and other crops.'

Only the lure of huge gold deposits that may bring in 10-20 tonnes annually for a decade or two makes mining companies persist in trying to overcome the combined hurdles of climate, lack of infrastructure and environmental worries, to say nothing of the political battles with governments naturally eager for their share of the boom. So, unlike the present surge in Australia, where new open-pit projects open within weeks, the drive for epithermal gold is a slow one, calling for persistence, patience and plenty of long-term capital. The full story will unfold only over the next fifteen to twenty years, so that going into the next century the rim of fire will contribute a good share of the world's gold. 'I believe we could see 200 tonnes annually,' David Tyrwhitt of Newmont Mining's Australian division told me.

Epithermals in Indonesia

Newmont themselves are busy exploring the possibilities of the next stage of epithermal development in Indonesia, where the islands of Kalimantan (formerly Borneo), Sumatra and Sulawese (formerly Celebes) all show great promise. Already there are alluvial gold rushes on these islands as local people scout out the high-grade rock where the epithermals break the surface. On the wall of his Melbourne office, David Tyrwhitt showed me a huge colour photograph he has taken on Kalimantan of a shanty town where five thousand or more local diggers are already extracting gold from the Masuperia deposit. Using water monitors and wooden sluice boxes, they are getting grades of 20-40 grams a tonne. 'They've even gone underground, following the main vein to depths of fifty or sixty metres,' Tyrwhitt said. 'They break up the rocks underground by heating them with fire and throwing on cold water.' One problem facing mine developers at Masuperia (where BP Minerals is working with the Indonesian government's mining department, Aneka Tambang) is how to clear these individual gold diggers and get going with a proper mine. If they can, the rewards look substantial. Some veins are several metres thick, with grades of 40-60 grams. Besides the main vein system, there is a large underground mushroom-like cap thought to contain up to 100 million tonnes of ore grading up to 3 grams. BP Minerals hope they have a winner, but cannily say 'it is early exploration days'. Masuperia probably contains between 60-150 tonnes of gold and, if the hazards of operating in a remote corner of Indonesia can be overcome economically, Masuperia could become a major mine some time in the 1990s.

Indonesia offers other good prospects. The most immediate is an alluvial gold deposit at Kasongan on Kalimantan, run as a joint venture by two Australian companies, Pelsart Resources and Jason Mining, with local Indonesian partners. Kasongan starts up in 1987 at 3-4 tonnes annually. On a grander scale, the Kelian project on Kalimantan, with Australia's CRA as main shareholder, has initial reserves of 27 million tonnes grading at 2.2 grams. The mine is due to start in 1989 with initial output at about 6 tonnes annually. Another good candidate is nearby at Mount Muro, which is being explored by Duval, Pelsart Resources and Jason Mining. A host of local diggers is already working the surface area, sometimes getting recovery grades of 30 grams. Test drilling has indicated isolated grades as high as 140 grams, with an additional 4,500 grams per tonne of silver. Mount Muro could become a mine in the 1990s.

This is the time-scale for Indonesia. The bonanza is not next year, but through the 1990s the country can develop to yield at least 50 tonnes, perhaps even 100 tonnes, of gold a year. Australian entrepreneurs, like Perth's Kevin Parry whose Pelsart Resources is busy with no less than nine prospects on Kalimantan and two on Sumatra, have grand ambitions that Indonesia will produce 150-200 tonnes. Other mining companies, with longer experience in the obstacles to be overcome in this region, take a more cautious view. No one denies that Indonesia can become a significant player. A special map prepared by the Sydney stockbrokers Jacksons for the magazine *Australian Business* in 1987 listed no less than thirty-two possible gold deposits in Indonesia scattered right through this vast archipelago from northern Sumatra three thousand miles to West Irian, the huge island split between Indonesia and Papua New Guinea.[1]

Unveiling the Under-explored

Looking at this map, which embraced the whole rim of fire, reminded me of a remark by Don Morley, finance director of Western Mining in Australia, who pointed out, 'The area is, by world standards, grossly under-explored.' So I counted up what else the map billed as potential gold mines on the long swing out from Indonesia into the Pacific. Papua New Guinea has thirty-three current

---

1 'Rim of Fire', *Australian Business, 1 April 1987*

mines or prospects, the Solomon Islands have one small operation and fifteen possibles, Fiji has the Emperor and twelve other prospects, Vanuatu (formerly New Hebrides) has seven prospects and New Zealand has twelve. Now, as Mark Twain observed, 'A gold miner is a liar standing beside a hole in the ground.' Not all the prospects will mature. Morley added the caveat that 'the quality of a prospect may have to be substantially higher than it need be elsewhere in order to justify the investment risk'. But the very fact that no less than forty-five mining groups, including some big guns like Australian Consolidated Minerals, Broken Hill Proprietory, BP Minerals, CRA, Newmont, Pancontinental, Placer Pacific and Renison Gold Fields, are out with their drilling rigs underlines the magnetism of the rim of fire. Maybe it is under-explored now, but the explorers are moving in.

Obviously Papua New Guinea offers the best potential. The successful Bougainville and Ok Tedi copper-gold projects have already got the politicians and people more accustomed to dealings with mining companies. The politicians have learned to get a share for the government in each project, and to insist not only on environmental controls but that mining companies stay with projects that look like going sour. Ok Tedi, for instance, is essentially a gold 'cap' over a huge copper deposit. With copper prices weak, there was a temptation to mine out the gold cap quickly by 1988 and pack up. The government demanded a long-term commitment to keep going on copper, once the best of the gold was gone. But as Bougainville and Ok Tedi pass their best gold years, so Misima, Porgera and Lihir Island will take over. Consequently Papua New Guinea's output will remain at least 40-50 tonnes annually through the 1990s, with a chance of going rather higher if other prospects pay off.

The Philippines, too, will remain a substantial producer, contributing at least 35-40 tonnes a year from the combination of existing epithermal and copper-gold mines, supplemented by extensive alluvial mining by thousands of prospectors. At the most conservative estimate, these prospectors are finding 12-15 tonnes annually from twenty-two sites, mainly in Luzon, Visayas and Mindanao. The richest pickings have been around Mount Diwata in Davao del Norte province in southern Mindanao, where at least 35,000 panners have their camps. Most miners manage to earn 60-100 pesos a day ($3-5), which is a good improvement on the 15 pesos a day they might hope to earn from farming or government work. In the more formal mining sector, Benguet, the leading mining group whose former president Jaime Ongpin became Mrs Aquino's minister of finance, has successfully cut its costs to enable it to keep open. 'We've trimmed the fat,' Delfin Lazaro, Benguet's new president, told me.

The real test in the Philippines is whether foreign mining companies (mostly Australian), who held back from investment during the final years of the

Marcos regime, will pluck up their courage to back President Aquino by going ahead on nearly a score of porphyry copper-gold and low-grade epithermal deposits that have been pinpointed. Several good epithermals have been identified both on Cebu island and in northern Luzon close to established gold-producing areas. The problems of infrastructure, therefore, are not as great as in Papua New Guinea or Indonesia. The hazard is largely political risk, both in terms of the stability of the new administration and the danger that many of the proposed mines are in, in areas controlled by communist guerillas. The encouraging sign is that Jaime Ongpin has moved from Benguet to the Ministry of Finance, and should be able to shape the Aquino government's attitude towards foreign investment in mining.

Among other patterns of Pacific rim islands, one of the best bets is Waihi on the north island of New Zealand. Waihi is a class epithermal deposit that was mined for many years earlier this century, yielding nearly 250 tonnes of gold. Now AMAX, in conjunction with Australian Consolidated Minerals (ACM) are proving up an open pit which promises 15 million tonnes of 3.1 gram gold and 30 grams of silver. They are opposed, however, by a strong environmental lobby which does not want the beautiful Coromandel peninsular spoiled by this mine, or another potential epithermal mine at Golden Cross a few kilometres away. The mining companies, who believe that the Coromandel peninsula is one of the best regions in the entire South Pacific for epithermal gold, have sought to reassure both the government and local people that every care will be taken. The New Zealand government remains divided on the issue, and the go-ahead has been severely delayed. Miners feel confident, however, that they will win the day. 'The environmental controls are easing, and Waihi will be open by the early 1990s at 2 tonnes a year,' predicted Ken Fletcher of ACM. 'New Zealand as a whole could be producing 15 tonnes annually by the end of the century.'

In the Solomon Islands, where output so far is limited to a few thousand ounces of alluvial gold, formal mining is in its infancy. However, exploration in central Guadalcanal by Arimco Minerals and Cyprus Minerals has revealed a huge low-grade epithermal deposit of over 50 million tonnes at 2 grams, known as Gold Ridge. The project could reach production late in 1988 or 1989, although expansion in the Solomons will depend on the government's willingness to welcome mining companies whose presence could radically alter the way of life and simple economy of these Pacific islands. No significant output is likely before the late 1990s.

The same is true in Vanuatu, where Australian mining groups, including City Resources, Hill Minerals, Jason Mining and Paragon Resources, are prospecting scores of possible sites. Patience is needed. 'We're looking, but we haven't found anything good yet,' admitted Stan Lewis of Paragon.

For all their promise, epithermal gold deposits are still something of a geological dilemma. 'The epithermal is a tricky animal,' said Geoff Hopkins, the chief geologist of Western Mining, which has a twenty per cent stake in Fiji's long-running epithermal mine, Emperor. 'There's a lot of indications; you follow them, but it doesn't get you anywhere. We are at a very early stage of the exploration cycle.'

In that perspective, the epithermals of the Pacific Basin will not revolutionise gold output overnight, nor even in a decade. Instead, as they are gradually understood and exploited, they will provide a regular new return for gold mining well into the twenty-first century.

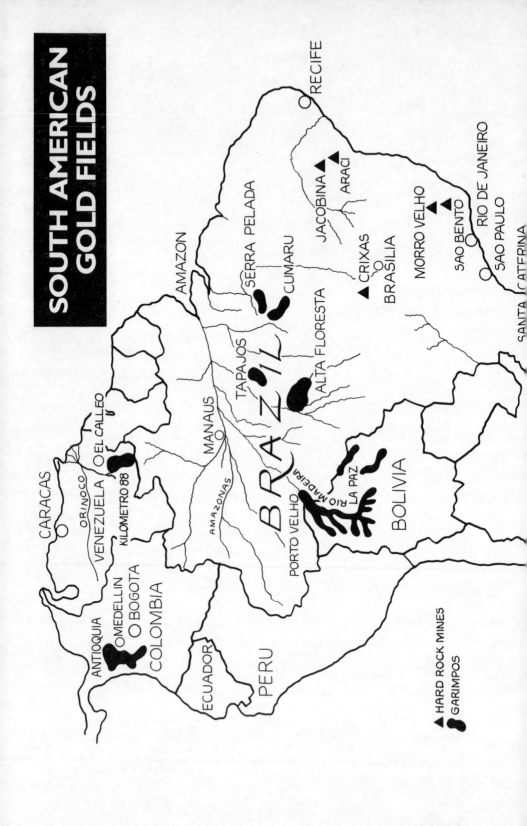

SOUTH AMERICAN GOLD FIELDS

CARACAS

ORINOCO

VENEZUELA

EL CALLEO

KILOMETRO 88

ANTIOQUIA

MEDELLIN

BOGOTA

COLOMBIA

ECUADOR

PERU

AMAZON

MANAUS

AMAZONAS

SERRA PELADA

CUMARU

TAPAJOS

ALTA FLORESTA

BRAZIL

PORTO VELHO

RIO MADEIRA

LA PAZ

BOLIVIA

CRIXAS

BRASILIA

JACOBINA

ARACI

MORRO VELHO

SAO BENTO

RECIFE

RIO DE JANEIRO

SAO PAULO

SANTA CATERINA

▲ HARD ROCK MINES

GARIMPOS

Chapter 7

LATIN AMERICA:
SEEKING NEW SERRA PELADAS

The miners of Serra Pelada personify the re-birth of mining in Latin America in the 1980s. Gaunt, grey figures labouring through mud with burdens of gold-rich dirt have haunted magazine pages and television screens for some time. Forty thousand men in soaked grubby shirts and shorts burrowing out a great pit in the Brazilian wilderness 270 miles south of Belem on the Amazon delta have revealed the harshness, the sheer slog of a gold rush. Their triumphs have hit headlines. The lucky few found nuggets of unparalleled size. One *garimpeiro* (as prospectors are called) turned up a 6.7 kilo nugget one morning and sold it for $84,000. Many made a better income than they could possibly have done as farmers or struggling to find work in Brazil's overcrowded cities.

Serra Pelada has yielded nearly 60 tonnes of gold since its discovery in 1980. But it has been a dangerous way to make a living. As the *garimpeiros* dug deeper, the sides of the hole often collapsed, burying them in mudslides (thirteen died in a single cave-in in October 1986). Such hazards were little discouragement. Driven by poverty and the hope of sudden wealth, close to half a million *garimpeiros* have trekked to Serra Pelada, Cumaru, Alta Floresta, Amapa and other alluvial outcrops throughout Brazil. Perhaps another half a million have joined alluvial gold rushes in Bolivia, Colombia, Guyana, Peru and Venezuela, often encouraged by rapidly depreciating currencies that sent local gold prices soaring. Working with shovels and water jets on land and with suction pumps attached to small boats in the rivers, their sheer hard grind has lifted annual Latin American production from around 50 to well over 150 tonnes in less than ten years. On every visit to Latin America, I hear tales of new rushes. In 1985 it was to the Rio Madeira on the Brazil-Bolivia border; in 1986 to Tucuman on the Rio Fresco, a tributary of the Amazon, and to El Dorado and Kilometro 88 in southern Venezuela.

Brazil has led the way. Judging the precise quantity scoured out by the *garimpeiros* is not easy. Output is certainly·up from around 10 tonnes a year to comfortably over 60 tonnes. Brazil's Banco Central alone has been buying 20-30 tonnes most years through buying offices at the *garimpos* (as the actual alluvial sites are known), while close to 30 tonnes has been arriving clandestinely in gold dealers' offices in New York and Miami.

The lesson of most gold rushes, however, is that the bonanza lasts from three to eight years. By then the richest surface deposits have been depleted, and the remaining reserves are no longer easily accessible to the individual prospector. Mechanisation, or even underground mining, is then essential. Hundreds of small concessions become concentrated in the hands of a few 'owners', who work over the tailings and try to bulldoze back the sides of a pit. Production inevitably falls off. Brazil is now at that stage. In its best years Serra Pelada alone gave up 13 tonnes of gold; in 1986 it was only 3 tonnes. 'The days of Serra Pelada *garimpo* are numbered except for possible retreatment of tailings,' reported Peter Rich of Desenvolvimento Aluvional, who is one of the foremost authorities on alluvial mining in Latin America. The story is the same at other *garimpos* on the Rio Tocantins, Rio Piranhas and Rio Gropori in the Tapajos region of Amazonia. Although 60,000 men are still working there, buyers in the town of Itaituba, where most go to sell their gold dust, noted purchases were down twenty to thirty per cent during 1986. Rich calculated that *garimpos* output fell back under 60 tonnes in 1986, and it could decline to 35 tonnes by about 1991. New discoveries may be made, especially in the vast Amazonas region, which would halt the fall, but the present picture is that new *garimpos* are not being found as fast as existing ones are being worked out.

The Big Guns Target Brazil

Brazil's gold boom, however, is not over. The character is changing. The trend for the remainder of this century will be a rapid expansion of primary gold mines by international mining groups. They are convinced that greenstone belts in Brazil's extensive archean shield host major gold deposits. 'Brazil contains a significant proportion of the last pre-Cambrian shield areas of the world which remain relatively unexplored,' notes Don Morley, finance director of Australia's Western Mining group, which has extended its test drilling to Brazil. 'Such areas have historically produced the world's major mineral fields in Canada, South Africa, Australia and Siberia.'[1] A clear sign of Brazil's untapped potential, Morley believes, can be derived just by comparing its present primary output with countries like Australia and Canada, which have similar archean geology. At present Brazil has less than one-third of the gold production of these countries, expressed in kilos of gold per square kilometre. 'It is unlikely', he argues, 'that the archean shield of Brazil will prove to be less prolific.'

An encouraging pattern is already emerging. Major mining houses like Anglo American and Gencor from South Africa, Rio Tinto Zinc and BP Minerals from Britain, and Cominco from Canada, along with Australia's Western Mining, are all active. Brazil's Mining Institute calculates $1.5 billion is being in-

1 Donald M. Morley, 'Other Free-World Gold Production', paper presented at Gold 100 Conference, Johannesburg, September 1986

vested in gold production between 1985 and 1990.

Anglo American led the way some years ago by taking over Brazil's oldest gold mining group, Mineracao Morro Velho, whose mine in Minas Gerais province started in 1835 and still produces a steady 3.8 tonnes annually. Anglo then developed the Jacobina mine in Bahia province, which is already yielding 1.2 tonnes and is forecast to double output by 1989. Much higher production comes from Anglo's expansion, with Brazilian partner Bozano Simonsen, of the Cuiaba and Raposas deposits close to Morro Velho, which are set to produce 5 tonnes annually. To complete its emerging Brazilian stable, Anglo bid aggressively to take over from Kennecott a fifty per cent share (with International Nickel as the other partner) in the proposed Crixas mine in Goias province. Crixas is viewed as one of Brazil's best prospects, with recoverable grades of 11 grams per tonne already confirmed and occasional much higher grades encountered as drilling proceeds at depth. The mine should come on stream in the early 1990s with an initial output of 4 tonnes annually, rising to 8 tonnes by 1995. If it lives up to expectations, Crixas will be the largest mine in Brazil and lift Anglo's overall production in the country close to 20 tonnes. Meanwhile Gencor, with Brazilian partner Amira Trading, has brought its Sao Bento mine in Minas Gerais province into full production in 1987, and has high hopes for another small project two miles away which could start up by 1989. Rio Tinto's Morro de Ouro Paracatu also began in 1987 and should yield 3-4 tonnes annually. Simultaneously, BP got into business with the small Sta Martha mine near Brazil's border with Bolivia, which has a proven grade of 15 grams per tonne and should produce 2 tonnes annually.

Hard on the heels of these international mining houses, several Brazilian groups have joined the rush. One aggressive newcomer, Companhia Mineracao et Participaoes (CMP), backed by Roberto Marinho who runs the TV Globo television network and investment banker Monteiro Aranha, swiftly raised $45 million in a jazzy launch on the Sao Paulo stock exchange. CMP holds concessions in four major gold mining areas, and already gets about 2 tonnes a year from various small operations. Another challenge comes from CVRD, the giant state-owned iron ore company, which is switching its exploration sights towards gold. It already has a small heap-leach operation at Araci in Bahia province, and is developing another open pit at Anta nearby. Totting up these and other small mines already proposed, close to 40 tonnes of gold will be produced by Brazil's gold mines by the early 1990s, compared with under 10 tonnes in 1986. If other mining groups like Cominco and Western Mining also come up with good prospects, mine output could easily clear 50 tonnes between 1995 and 2000. Assuming *garimpo* output stays around 30-40 tonnes, Brazil's total production will range between 75-90 tonnes. Taking an even longer view, the fact that much of the archean shield is still relatively unexplored suggests that, going into the next century,

Brazil's output could rise further. As Western Mining's Don Morley summed it up, 'Large numbers of greenstone belts and associated gold deposits remain to be discovered.'

The limelight on Brazil should not obscure the growth that has taken place elsewhere in Latin America. In Colombia and Peru production has more than doubled since 1980; in Chile it has tripled, and in Venezuela it has soared at least ten-fold. The Colombian expansion has been almost entirely due to alluvial operators pushing up output from around 10 tonnes annually to over 25 tonnes. They have been encouraged for several years by a substantial premium over the gold price paid by the central bank, Banco de la Republica, with the bank offering as much as thirty per cent over the gold price. The incentive to seek out gold not only spurred Colombian prospectors, but sucked in gold from many other alluvial deposits in neighbouring countries. In Colombia up to 200,000 barraqueros (as prospectors are called here) have joined the search for gold. Spreading out from frontier towns like El Bagre in the Antioquia province of north-western Colombia, they have worked their way up the Nechi and Saldana rivers, where the big dredges of Mineros de Antioquia and Mineros del Choco also operate. Mineros de Antioquia's six dredges, gobbling up the gravels of the river bed, produce about 130 kilos monthly, but this more formal mining accounts only for a small part of Colombia's gold. The barraqueros shoulder the main burden, with hydraulic monitors and excavators to turn over the river banks, and small boats fitted with pumps to scavenge the river beds. They are leaving a devastated landscape devoid of vegetation behind them, and the mercury they use to separate out the gold is polluting the rivers. But, as in Brazil, the best years may be over. 'It's reached the peak, all the easily accessible material has been mined out,' a geologist just back from Colombia told me.

The same is also true in Chile, where the spectacular epithermal El Indio mine, launched in 1980 with grades as high as 358 grams per tonne (plus 1,000 grams of silver), will soon have depleted its richest reserves. The mine still yields around 9 tonnes a year, but this will fall away to about 6 tonnes by 1990. The Chilean government has sought to counter this decline by setting up Plan Aurifero Nacional to stimulate production at small alluvial prospects, but this is not achieving significant growth. A better hope is a revived silica deposit, Hueso de Silica, which is also rich in gold, that is being expanded by Codalco.

Opening Up Venezuela

The pace-setter today is Venezuela. Small-scale gold production has been going on in Venezuela for many years, notably around El Callao down towards the Guyana border. Ambitious plans in the early 1980s by the government mining group Minerven to expand output at El Callao met with limited success. Instead, a spate of new alluvial discoveries since 1984 has transformed the scene. The incentive for the gold rush came from the drastic devaluation of the Venezuelan bolivar after oil prices crashed. The local gold price shot up from 55 bolivars per gram to 255 bolivars inside three years. Suddenly gold prospecting became very profitable. Over 30,000 *mineros*, who previously prospected for alluvial diamonds in southern Venezuela, switched to gold. They were joined by thousands of new-comers, fanning out in boats along the Caroni and Paragi rivers, and ranging south on the lonely road from El Calleo towards the Brazilian border to such aptly named spots as El Dorado and Kilometro 88. By early 1987 at least 25,000 *mineros* were working thirty-two main concessions blocked out along the Caroni river just north of Kilometro 88. Results were often spectacular. One small dredge was reported to be getting 4 kilos a day, another as much as 5 kilos. The gold often turned up in exquisitely formed nuggets sometimes weighing 30 or 40 grams, with a purity of well over 900 fine. One dealer in Miami, where much of the gold is spirited for sale, showed me a rare assortment of nuggets he had as-sembled. They were too beautiful to melt down and refine, he said, and anyway worth far more as collectors' items than their simple gold content.

The difficult question in Venezuela is how much is being produced? The gold is dispersed to many destinations. From St Elena in the south of Venezuela it tends to go over the border into Brazil, often bartered for essential supplies like fuel for an outboard motor or tyres for a pick-up truck. A lot more was smuggled over the border to Bogota as long as Colombia's Banco de la Republica paid a high premium. The Venezuelan central bank, following their example, itself set up a gold-buying unit and was hoping to purchase 6 or 7 tonnes annually. Gold also flows out to Miami and Switzerland. Several dealers I talked with suggested that altogether output was already 25 tonnes annually, with more to come. Jaime Tugues, one of Venezuela's best-known dealers who runs the El Platino refinery in Caracas, told me he thought it was at least 18 tonnes in 1986.

Judging from the spate of American companies setting up office in Caracas and heading up country to sell everything from pumps to gravel concentrators, the rush is gathering momentum. 'It's still very inefficient at every stage,' Courtney Brewer of Inversiones High-Tech Mining told me. 'There's tremendous potential for improving technology and using good machinery. These are early days.' He was busy trying to market a machine known as The Knelson Concen-trator ('Turn your golden dreams into golden realities') to improve the recovery rate on gravels from the Venezuelan rivers. At the moment a great deal is simply being lost, and a fortune is still waiting to be retrieved in gravels dumped by the rivers after inefficient sluicing. The equipment salesmen are being followed by several small mining houses. Greenwich Resources, fresh from pioneering in the Sudan and Egypt, is scouting the interior. Discounting some of the more exagg-erated claims, it is probable that over the next decade Venezuela will produce 20-

30 tonnes annually, lifting itself into second place in the Latin American league.

The constant need to find new alluvial deposits to replace depleted old ones throughout Latin America does mean that overall output will not advance much from the present plateau around 170-190 tonnes. One country's advance will be counter-balanced by another's decline: Venezuela up, Colombia down. Even so, Latin America is now a significant contributor to world markets, where a decade ago a much lower level of production could be absorbed locally. This is particularly noticeable on any rise in the gold price, which brings a surge of home-refined Brazilian and Venezuelan gold bars - and nuggets - flooding into Miami and New York.

Chapter 8

ARABIA TO ZIMBABWE: TUTANKAMUN REVISITED

Anyone who has marvelled at the gold treasures buried with the boy-king Tutankamun, who ruled Egypt from 1361-1352 BC, must have given a passing thought, as they gazed upon his coffin of solid gold weighing over 100 kilos and the great mask of beaten gold that shrouded the head of his mummy, to the question of where did all that gold come from over 3,000 years ago? The answer is from the Red Sea hills and the Nubian desert, which are part of the greater Arabian shield whose greenstone belts match geologically with the pre-Cambrian shields of Australia, Canada, the Soviet Union and many other gold-producing areas. This was the cradle of gold mining. As the gold historian, C.H.V. Sutherland, observed, 'It was from Egypt that the ancient world as a whole learned the main principles of gold mining and metallurgy at a very early period.'[1] And in the first real account of mining in the second century BC, the writer Diodorus noted, 'On the confines of Egypt...and Ethiopia is a place which has many great mines of gold.'

Not much has changed. A report from the London stockbrokers James Capel in October 1986 on the re-opening of a mine at Gebeit in the Red Sea hills of Sudan, as a joint venture between the London-based Greenwich Resources and the Sudan government, remarked, 'The Gebeit mine is said to be one of the world's oldest mines and since the discovery of gold in the area around 1500 BC has been worked over three main periods.' The mine, which came on stream in 1987, contains some reserves grading as high as 35 grams per tonne and will produce around 2.5 tonnes annually for at least five years. Cash operating costs are forecast at an astonishingly cheap $85 per ounce, and capital costs will be recovered in under two years.

The revival at Gebeit by Greenwich Resources, a small company which prides itself on prospecting in countries neglected by mining giants, is only the beginning of plans to revisit old gold prospects on both sides of the Red Sea. Besides its deal in Sudan, which also embraces mining leases in the Nubian desert and the Nile valley, Greenwich has signed on with the Egyptian Geological Survey and Mining Authority for exploration at the El Sid concession close to the Red Sea port of Ouseir and in the desert at Barramiya between Luxor and the

1 C.H.V. Sutherland, *Gold*, Thames & Hudson, London, 1959, p.29

Red Sea. Some provisional drilling at Barramiya a few years ago hinted at a large orebody grading 2.1 grams.

Meanwhile, just across the Red Sea in Saudi Arabia, production is finally set to begin early in 1988 at the Mahd ad-Dhahab (Cradle of Gold) mine in the desert between Jeddah and Medina. The mine has been developed by the Saudi oil ministry Petromin, with assistance from Consolidated Gold Fields of London, and promises grades of 26 grams (with a bonus of 90 grams of silver) that could deliver up to 2 tonnes annually. Since other small orebodies close by can probably be developed and treated at the mine, Mahd ad-Dhahab could run for fifteen years. The Saudis have also announced the go-ahead on another gold project, in conjunction with Sweden's Boliden group, at Sukhairat in the north of the kingdom, that is forecast to yield 1.5 tonnes annually for about ten years from 1990.

Such projects are not going to radically transform world gold outlook, but they underline what we have already observed everywhere from Brazil to the Pacific rim of fire, that there are several relatively uncharted regions offering good potential for growth over the next ten to twenty years. In Saudi Arabia, for example, the focus has been so much on oil that it is only recently, with weak oil prices, that the government is more actively approving the search for other minerals. Finding more gold may not be easy. 'I'm sure there are other deposits, but a lot will be covered by sand or lava flows,' explained a geologist who worked for two years setting up Mahd ad-Dhahab. 'It's only in the hills that it's easy to explore. Elsewhere it's a very hostile environment.' He estimated Saudi might get production up to 5 tonnes a year, with a similar amount from the ancient mines of Sudan and Egypt.

Historical sources of gold are also being reassessed in West Africa, where Arab traders bartered for gold dust at least a thousand years ago. The Gold Coast, as it was called, was a major supplier to dealers in the London gold market in the eighteenth and nineteenth centuries and, until South African discoveries were made, actually accounted for one-third of world output. Even in 1961 the newly independent state of Ghana ranked fifth in the world league at 34 tonnes a year. Since then, production at the huge Ashanti gold mine one hundred miles north-east of Accra, which is still the biggest employer in the country, has declined steadily. Ashanti, in which the Ghanaian government has a fifty-five per cent stake and 'Tiny' Rowlands's Lonrho has forty-five per cent, simply could not get the foreign exchange allocation to bring in new equipment or develop new underground reserves. However, in 1985 the World Bank arranged new financing so that Ashanti could deepen two shafts and install a carbon-in-pulp treatment plant. These improvements will raise output by fifty per cent over five years up to around 12 tonnes annually. The World Bank is also funding a three-year programme to double the output of three smaller Ghanaian mines operated by the

State Gold Mining Corporation in the Tarkwa and Prestea gold fields. Without doubt, the potential in Ghana is considerable; it has high-grade deposits close to the surface. The obstacle has been the political risk, which made mining companies ultra-cautious about investment for fear that they might not be able to repatriate capital or profits. An ambitious scheme with tax honeymoons and exemptions, announced in 1981 to develop fourteen mines and lift output to 60 tonnes annually, came to nothing. But the current modest plans will at least increase annual production to 18-20 tonnes.

Greater confidence is being shown next door in the Ivory Coast, where quartz veins similar to those of Ghana's Prestea gold field have been identified. Canada's Eden Roc Mineral Corporation is working on two neighbouring deposits in the Afema sheer zone, which spans the border between the two countries, where some grades are as high as 15 grams per tonne, and recoverable grades average 6-7 grams. They plan to start production in 1988 with a potential yield of around 2 tonnes a year. The project has at least a ten-year life, and initial costs are a modest $200.

The potential for low-cost operations is also encouraging modest expansion in Zimbabwe, where gold mining has recovered from the years of civil war and risen to around 15 tonnes. Cluff Mineral Exploration has opened its small Royal Family gold mine at Filabushi, and sees other promising prospects with costs of less than $150 an ounce in the north of the country. The government itself encourages small miners because it helps employment in the rural economy, and is building a new gold refinery in Harare as an alternative to its long-standing dependence on South Africa's Rand Refinery. (Meanwhile Zimbabwe's gold is being refined at the Perth Mint in Western Australia.)

Monitoring African production precisely is not easy. Alluvial gold output has increased steadily during the 1980s. Many Africans, especially in Zaire, have forsaken work - and even school - to join the search for gold. Zaire is the best source, but Mali, Ivory Coast, Guinea, Senegal, Cameroon, Gambia, Togo, Tanzania and Ethiopia are all contributing. The gold is smuggled out to Belgium, Britain, Dubai and Switzerland, usually as dust or in rough bars of 920-930 fine. At one Swiss refinery I saw wooden boxes filled with a wonderful rich gold dust, like grains of fine sand from some superb beach. Close to 25 tonnes a year is filtering out of Africa, of which perhaps 10 tonnes is from Zaire, and 3-5 tonnes from Mali and Ivory Coast.

Zaire's gold tends to move through Burundi to catch flights to Europe from East Africa, while West African output usually collects with traders in Bamako, the capital of Mali, for flights to Brussels and Geneva. The priority, once couriers get the gold to Europe, is fast payment in cash. Refineries compete with each other to offer the quickest assaying, often picking up the gold from couriers at airports and paying within twenty-four hours. The couriers then use the money to buy jewellery, watches or other consumer goods to take home to trade. Some Geneva dealers have encouraged expansion of production by arranging for pumps and other simple pieces of mining equipment. But the flow of gold dust from Africa has eased slightly and, as in Latin America, it looks as if the best sites have been creamed off. Alluvial output can probably remain at 20-25 tonnes a year, but the future growth will be through small mining groups, such as Greenwich Resources or Eden Roc, venturing into Africa for serious exploration. Undoubtedly there is plenty of gold to be found.

Chapter 9

THE SOVIET UNION: GORBACHEV GOES FOR GOLD

The Soviet Union is the eternal enigma of the gold business. For over half a century both her production and sales to the West have been a closely guarded secret. Bullion dealers, mining houses and the CIA all make their estimates. What they know for certain is that the Russians are good professionals in gold. They remember, too, the famous essay written by Lenin in which he stressed that although gold was an unfortunate necessity in the capitalist present, it would be useless in a socialist society. Gold, Lenin predicted, would ultimately be used only to cover the walls and floors of public lavatories. But, until that happy day, the Soviet Union must play the gold game. 'Sell it at the highest price, buy goods with it at the lowest price,' he wrote. 'When living among wolves, howl like the wolves.'[1]

That phrase sums up Russian gold operations for the last generation. In the gold market they 'howl like the wolves' - they are good capitalists. Analysts sometimes suggest the Russians dump gold. Why should they? It is a key foreign exchange earner, and they always seek the best price. The Bank for Foreign Trade in Moscow, which handles gold operations, has become an astute trader, using forward sales, futures and options. They will sell on Comex in New York and undo with EFPs (Exchange for Physicals) in Zurich. The Bank will often not only stay away from a weak market, but buy back to support a falling one. They are an integral part of the international gold scene. As major gold producers, it is just as natural for the Soviet Union to sell gold as it is for South Africa. They must always be viewed in that light.

In the close world of gold, actual sales can be tracked fairly accurately. The best estimates suggest Soviet sales fluctuated between 80 tonnes and 200 tonnes annually from 1980 to 1985, picked up closer to 280 tonnes in 1986, and are likely to remain on a higher level of 250-350 tonnes in the years ahead. The precise amount of sales each year is determined by the Soviet Union's foreign exchange needs. Oil, natural gas and arms sales are her best foreign exchange earners; gold is more a balancing item on the bottom line. If Russia is in surplus, then gold sales are low; if in deficit, because of a poor grain harvest or, as in 1986, lower oil prices, then sales are higher. The level of sales is also not necessarily pegged to the

1 Lenin, 'The Importance of Gold Now and After the Complete Victory of Socialism', 1921

dollar price of gold; much of the Soviet Union's imports are billed in deutschmarks, so her eye is more on DM earnings than dollars. Thus a higher dollar price does not mean less gold sales to achieve the foreign exchange required, unless the DM price has improved equally.

While foreign exchange requirements are the dictator of gold sales, bullion dealers also detect a more pragmatic attitude to gold sales since Mikhail Gorbachev came to power. He appears to regard gold as just another commodity useful for earning foreign currency, like oil or diamonds or furs, which the Soviet Union sells regularly; it is not something to be husbanded in reserves, especially when his plans to revive the Russian economy call for importing expensive technical knowledge. The Soviet Union therefore will sell all production that is surplus to her own demand for electronics or other specialist uses, and to meet some commitments to her Comincon partners. This local offtake is about 70-90 tonnes a year, according to Western estimates, leaving the balance available for sale.

The question is, how much does that leave? What is the level of Soviet production and, more important, how fast is it growing? Here the mystery, despite Mr Gorbachev's policy of·glasnost (openness), remains. Some clues, however, exist. After I gave an informal talk on Soviet gold production and sales in 1986, a regular visitor to Moscow came up to me and suggested politely that I had got only one thing wrong. I was under-estimating how steadily Soviet output was growing in the late 1980s. He reckoned that output was up by ten per cent in 1986 alone, and that from the level of 300-330 tonnes generally assumed in recent years it would rise towards 400 tonnes before the end of this decade. That insight may be correct. Soviet production is growing, just as it is in the Western world and in China. Indeed, the fact that China's output (see Chapter 10) is expanding rapidly, especially close to her north-western borders with the Soviet Union, is a signpost that the Russians must be making similar discoveries across the frontier. Moreover, Mongolia, which is theoretically an independent republic under the Soviet wing but has always sold its limited gold production through Moscow, has scouted out the possibilities of direct gold sales in the West, explaining that output is up.

In fitting the jigsaw puzzle together, a little history is useful. The Russians can look back on a long history in gold mining. Jason and his Argonauts, setting out about 1,300 BC in search of the Golden Fleece, were actually crossing the Black Sea in search of gold in the rivers of what is now Soviet Armenia. Alexander the Great conquered Armenia in 33 BC to secure gold for his vast empire. The modern era of Russian gold mining began in 1744 with the discovery of a quartz outcrop in the Ural Mountains, near the city of Ekaterinburg, which was mined regularly for the next forty years. Later more discoveries were made in the

Ekaterinburg area and farther east in the Atlas Mountains and down the upper tributaries of the Yenesi river, so that prior to the Californian gold rush in 1848 Russia was the foremost producer, yielding sixty per cent of all the world's gold. Although the Californian and Australian gold rushes rather eclipsed Russia, production continued to rise to about 45 tonnes annually by the mid-1880s (as South Africa started up), and 60 tonnes by 1914.

The real drive to expand production, however, came under Joseph Stalin in the late 1920s. He was concerned at Japan's potential threat to Russia's Far Eastern provinces, and determined to encourage migration to Siberia in an attempt to stimulate the economy there. He conceived the idea (nourished, so it is said, by learning how the California gold rush opened up America's West) of a similar flood of gold prospectors to Siberia. He created a Glavzoloto or Gold Trust to oversee the expansion, and even approved the hiring of American mining engineers to help.[2] One of those engineers later predicted the Soviet Union would overhaul South Africa by 1940 as the foremost producer. No one knows for sure, because Stalin made gold output a state secret. Indeed, it remains a criminal offence to divulge figures on productive capacity, production plans or targets achieved for all precious metals.

The basic structure of the gold mining industry set up by Stalin survives. Within the Ministry of Non-Ferrous Metallurgy, the Gold Trust, Glavzoloto, presides over fourteen regional *zolotos*. Lenzoloto, for instance, supervises the alluvial deposits along the Lena river in Siberia. Armzoloto watches over the reworking of the historic deposits in Soviet Armenia, including a major mine at Zod Pass, where a highgrade deposit has been worked for the last twenty years. The Zod mine is estimated to have a maximum output of 10 tonnes a year.

Zod, initially opened in 1966, marked a significant shift away from the alluvial gold fields of Siberia, which had been the source of up to two-thirds of output for many years. The balance shifted further in 1969 with the opening of the Muruntau mine in the central Asian province of Uzbekistan, between the Aral Sea and the Afghanistan border. The mine is in the heart of the Kyzyl-Kum desert, where a new town, appropriately called Zarafshan ('golden'), has been build to service it. From the start Muruntau was billed as a world-class mine, with reserves that will last well into the twenty-first century. The three main low-grade orebodies that have been located at Muruntau are estimated to contain between them over 1,000 tonnes of recoverable gold. The mine's shape is unusual. It is like a huge tree, with the trunk coming up from the centre of the earth, and the branches then spilling out close to the surface. The gold is extracted by a huge open pit that is already over 150 metres deep and is expected eventually to plunge to 400-450 metres, effectively lopping off the branches of the tree. Ultimately an underground operation may dig out the 'trunk'.

2 For a fuller account of this period, see Timothy Green, *The New World Of Gold*, Weidenfeld & Nicholson, London, Walker & Co., New York, 1981, revised 1985, Chapter 3

Assessing Muruntau's annual production by analysing articles about it in Soviet technical journals, and studying satellite photographs of the growing pit and waste dumps, has produced an astonishing range of estimates. The highest proposal is 143 tonnes, which would make it easily the largest gold mine in the world.[3] Consolidated Gold Fields, the London mining house, after a careful study of Soviet gold, concluded Muruntau was producing 80 tonnes a year.[4] More recently, however, estimates of Muruntau's output have been considerably scaled down. In an article in *Mining Magazine*, V.V. Strishkov, a consultant on Soviet precious metals living in the United States, concluded that at best Muruntau had a design capacity to produce 22 tonnes a year, and in practice achieved about 20 tonnes.[5] He reckoned that from start-up in 1969 through to 1985 Muruntau produced overall 213 tonnes of gold, and another 844 tonnes remained to be mined. On the assumption that output stayed around 20 tonnes annually, this gave Muruntau a life of at least forty years until 2026, and that was without going underground. This revision for Muruntau makes considerable sense. It is a low-grade deposit, with an average grade of 2.5 grams per tonne, probably declining as time goes on to 2.2 grams. To achieve 80 tonnes, let alone 143 tonnes, annually from such a low grade would require a level of mining and milling not achieved at any other gold mine. South African mines like Vaal Reefs or Driefontein Consolidated, which touch close to 80 tonnes a year, have much higher grades and are technologically far superior. Muruntau, in fact, seems more akin in scale to some of the Pacific's 'rim-of-fire' mines like Porgera, Misima Island or Lihir Island (although it is not epithermal), which have a potential output of 10-30 tonnes annually.

Given this more realistic dimension, it is prudent to keep overall estimates of Soviet production on the low side, although obviously Muruntau is a long runner. But, if it yields only 20, not 80, tonnes, then considerable leeway has to be made up in other production. The implication is that the Soviet Union is still more dependent on the alluvial production from its traditional Siberian sources of the Lena and Aldan rivers, the province of Magadan, and the Kamchatka and Chukotsk peninsulas on the Bering Sea opposite Alaska. The constraints of climate here are formidable. Mining is possible only during the summer months. Even then the permafrost often has to be thawed out. This means, for example, that the Russians cannot benefit greatly from new techniques such as heap-leaching, which has made so many new mines viable out West in the United States. The leaching dumps would freeze solid for much of the year. The Siberian placer deposits have also been filleted of their richest pickings. The future, therefore, has to depend on new discoveries in other regions.

The Russians have shifted the focus of their exploration to Armenia, Uzbekistan, where Muruntau is located, and into the neighbouring provinces of

3 J. Krason, *Mining Engineering*, November 1984, pp. 1, 549   4 *Gold 1978*, Consolidated Gold Fields, London

5 V.V. Strishkov, 'The Muruntau Gold Complex', *Mining Magazine*, September 1986, pp. 207-9

Kazakhstan and Tadzhikstan. Here in the mountains beyond the cities of Samarkand and Tashkent, they have apparently come up with many new deposits which can rival, if not outstrip, the Siberian output. The expansion, as it happens, matches closely with activity in China, where the autonomous region of Xingjiang just over the border has ambitious plans to expand output. Indeed, the signs coming out of the remote heartland of central Asia, where the Soviet Union, Mongolia and China all come together, is that plenty of gold is in prospect.

The test in all three countries will be how effectively it can be developed. Observers of Soviet mining argue that one of the reasons why gold output has stagnated at around 300-330 tonnes for a number of years has been the sheer inefficiency of both precious metal and base metal mining (from which about 60 tonnes of gold comes as by-product). Much will depend on Mikhail Gorbachev's drive to revive productivity throughout the Soviet economy. It is not just whether the mines themselves operate properly, but whether they can get the drilling rigs, mechanical shovels, trucks and bulldozers - together with spare parts - that they need. Muruntau, for example, requires over 200 trucks and thirty bulldozers.[6] A mining engineer in London told me, 'You tell me how well organised the Russians will become, and I'll give you a better idea of their output.' A hint of their problems in maintaining high standards has come from the fact that in recent years some of the 9999 good delivery bars from their refinery have not been quite up to title. Western refining experts, invited to Moscow to advise, found that although equipment was modern, maintenance was poor so that consistent quality was jeopardised. The very fact the Russians asked for help, however, signals a fresh approach. Their determination to improve their gold industry is also demonstrated by their central bank, which has started to finance the development of new mines through gold loans from its own reserve. This gold is then sold forward as part of the Soviet Union's regular sales programme, providing advance cash flow for the new projects (just as is happening in Australia and the United States). Effectively this means the Soviet Union is selling reserves to underwrite gold mine expansion, although the mines must repay the central bank in gold once they are on stream.

The true size of Soviet reserves is another guessing game. Usually they are calculated at between 2,000 and 3,000 tonnes, by estimating production since the 1920s and netting off sales. However, that is arbitrary, because how accurate are production guess-times? Some close observers of the Soviet precious metal scene believe the reserves are actually just under 2,000 tonnes, because production has been exaggerated. Whatever the precise amount, the Soviet reserves are by comparison rather similar to those of France, Italy, Switzerland and West Germany, and about one-quarter those of the United States. That is not much cushion as an

6 Ibid

emergency stock for a super-power. So, although under Mr Gorbachev Russia may use her reserves through loans to help boost output, a real run-down is unlikely. Rather, the sales of most-current production must be anticipated.

Despite the uncertainty about the precise annual output, no one disputes that the Soviet Union is comfortably in second place in the gold production league. One intelligence estimate put it at 325 tonnes for 1986, over 200 tonnes ahead of Canada and the United States. If growth is ten per cent a year, as is hinted to some Moscow visitors, then a target of 400 tonnes by 1990 is attainable, especially if productivity improves. Looking further ahead, the crucial factor will be if enough new discoveries can be made to balance the inevitable decline from the Siberian placer deposits as they are worked out. The necessity of replacing depleted ore reserves is one reason not to look for the kind of boom in Soviet output we are witnessing elsewhere. The Soviets' aim must be to lift output to, say, 400 tonnes and sustain that level.

The implication for the international market is that the Russians will have an excess over domestic or Comincon requirements of at least 250 tonnes a year, and perhaps up to 330 tonnes by 1990. They could also cut back on domestic needs if their electronics applications became more effective, and on sales of Comincon partners. The 'hard-core' domestic demand might be honed down to 40 tonnes, leaving close to 300-350 tonnes available for export sales. This is a much larger quantity than we have seen in recent years.

However, the price may not be unduly depressed. The Russians play the market very well. They sell into strength, and try to spread their sales in a reasonable pattern over the year. Both in 1986 and 1987 they took advantage of higher prices in the opening months to dispose of close to 100 tonnes, and then sat back for a while to await price developments, in the knowledge that they already had in hand over $1 billion of their perceived target of about $3 billion in gold earnings for the year. The Bank for Foreign Trade in Moscow, which has handled gold sales since an unfortunate debacle at the Wozchod Handelsbank in Zurich in the early 1980s, also has a bright team of dealers adept at using forward sales, futures and options to maximise their own profit and to create a smoke-screen on their actual net sales.

The long-term view of those sales must be that the Russians will remain as the market's second-biggest supplier. They will also be careful not to disrupt it. As a Soviet banker handling gold sales told me some years ago, 'We have no interest in cutting down the tree.' The actual amount of sales each year will depend on foreign exchange requirements. If oil prices stabilise and strengthen, then gold sales will be less. Moreover, there is a practical limit to how much extra gold can be sold in a year. If the Russians sold too heavily from reserves, then the price might collapse. They have the leeway to sell perhaps an extra 100-150 tonnes in a bad year, earning themselves another $2 billion or so, but beyond that sales would be counter-productive. In short, the Russians see gold as a useful way of earning $2-3 billion annually under normal circumstances, with a ceiling of $4-5 billion if they are in deficit. Watch the Soviet balance of payments to judge whether it will be the lower or the higher figure.

Chapter 10

CHINA: GOING FOR KAM

For the Chinese people, *kam*, gold, has long been appreciated as a symbol of wealth and store of value. *Kam yuk moon tong*, 'Gold and jade is everywhere in the hall', and *Kam ngan moon oak*, 'Gold and silver is everywhere in the house', are traditional greetings at Chinese New Year. While the proverb *Jan kam butt par hung lo for*, 'Real gold is not afraid of the fire of a real furnace', acknowledges its unique properties.

Faith in gold has clearly survived the Communist Revolution in the People's Republic of China. The recent 'open door' policy, designed to revitalise the antiquated economy, has spilled over into a drive to increase gold production. The initiative is not limited to state-owned mines. Individual prospecting has been approved since 1978, and is being further stimulated in a current drive to raise production between fourteen and sixteen per cent annually. 'The policy of encouraging collectives and individuals to mine gold will continue,' said Huang Yuheng, general manager of the China Gold Company of the Ministry of Metallurgical Industries, announcing new targets for 1990. 'More flexible policies will be adopted to further attract people to mine gold.'

Already well over 200,000 farmers in the countryside have forsaken plough shares for gold pans. Output rose, according to official figures, by sixty-three per cent between 1980 and 1985 and is forecast to double by 1990. The Chinese do not say precisely how much they mine each year, but it is estimated at 70-80 tonnes for 1987, and should top 100 tonnes by 1990.

The expansion has two motives. The primary one is the need for foreign exchange. 'Gold is foreign exchange,' a senior official of the People's Bank of China put it simply in Beijing. But he hastened to add that as China's living standards improved, so did the local demand for gold jewellery. 'We need lots for rings and chains,' he said. Gold jewellery sales were banned in China for more than twenty years after the Revolution, but have been permitted again since 1982. Already they are worth, at retail level, over $500 million annually, according to official figures. Such sales could absorb 20-25 tonnes annually, depending on mark-up and sales taxes. That is a very modest beginning. A Hong Kong gold dealer, who was also in Beijing during my own visit, suggested, 'As China's GNP grows, more and more of the population of one billion will resume the tradition of buying and hoarding gold.'

The serious intent of China's entry into both gold mining and the international gold business is in no doubt. The Bank of China, the commercial arm of the People's Bank of China, which handles overseas operations on behalf of the central bank, proudly showed me their new gold trading desk in Beijing, complete with glowing Reuters screen and back-up telexes. Their gold dealer was just off to Hong Kong for training. Already Bank of China in London and its sister bank Po Seng in Hong Kong are active participants in international gold trading. China also successfully launched in 1982 its own gold bullion coin, the Panda, selling in units from 1 ounce to 1/20th ounce. The Panda is one of the most beautifully finished low-premium coins (it is actually half proof quality). The design is changed each year to make it a more attractive collectors' item, and this has helped to carve out a loyal following in the United States, where nearly two-thirds of the coins are sold. The coins dated 1982 and 1983 now command a substantial premium, thus appealing increasingly to investors. The coin has been in such demand that both in 1986 and 1987 orders exceeded the capacity of the China Mint in Beijing. 'We are getting orders for 500,000 coins a year, but our capacity is only 300,000 at present,' Hu Ji Zhou, deputy manager of the China Mint Company, explained. The Mint has responded by raising its premium slightly, to 3.5 per cent on 1-ounce coins, and six per cent on ½-ounce coins. A tricky decision in the highly competitive bullion coin market, but justified because the Panda has a legion of regular collectors and weathers the fluctuation of the international market better than most bullion coins.

The need for gold for Pandas, for local jewellery sales and as a natural source of foreign exchange explains the Chinese government's decision to embark on a considerable investment programme in mining in the late 1980s. Under the direction of the China Gold Company (also known as the Gold Bureau) at the Ministry of Metallurgical Industries, close to $100 million is now being spent annually on modernising existing mines, with an extra $30 million allocated for exploration each year by the Geological Institute.

Actually, gold mining has a long history in China. The Jao Yuan mine in Shandong province has been worked, on and off, for over a thousand years. The small mines around Yiman, also in Shandong, date from 1655. These and other mines in Shandong, the eastern province of China between Beijing and Shanghai, are mostly small scale, often operated on a commune basis. In the last decade, however, a determined effort has been made to modernise the Shandong mines and open up new orebodies. The Xin-Cheng mine, close to the old Juan Yuan workings, started in 1979 and has a modest output of about 1.4 tonnes annually, from a grade averaging 9 grams. Three miles away, the new Jiao Jin mine has even better grades at 13 grams per tonne, and probably yields nearly 3 tonnes a year - large by China's standards. The newest Shandong mine, Hiaojiashi, which came on stream in 1987, is being billed as China's biggest pure gold mine. The mine is planned to produce 1,500 tonnes a day, but no grade is listed. Assuming the grade is similar to other Shandong mines, at 6-10 grams per tonne, Jiaojiashi could contribute perhaps 4 tonnes annually. However, it will not match the by-product output of gold from the Dexing copper mine, which opened in 1983 and is estimated to yield a bonus of 17 tonnes of gold annually.

The search for gold is not limited to the known goldfields of Shandong. Gold is now being mined in most of China's twenty-five provinces and autonomous regions, but is concentrated, besides Shandong, in Hebri and Liaoning in the east, in Heilongjiang in the far north-east near the Soviet border, and in Xinjiang in the north-west, close also to the Soviet goldfields of Turkestan. An official account, released in Beijing in 1986, reported gold was being produced in 429 counties of China, of which twenty-one counties each contributed over 10,000 ounces (311 kg) a year.

Reviving Private Enterprise

Clearly the major force in recent years has been the individual panner. Since 1978 China has urged local governments, collectives and individuals to run small gold operations in areas where large-scale mining would be unsuitable or uneconomic. They have even been encouraged to work on the fringes of state mining areas or on low-grade quartz veins in state mines. New gold camps have flourished. In one 3-mile long valley in the north-western province of Gansu, more than 20,000 diggers are scouring over thirty alluvial gold pits. They recovered almost 20,000 ounces (622 kg) in 1985, according to official reports. All told, at least 200,000 prospectors are active throughout China, and they are said to account for half of all production.

Individuals and collective mines should be licensed, and they are required to sell their gold to the government. The trouble is the price offered through the People's Bank of China is not so generous. Although the Bank raised its buying price from 697 yuan (US$188) an ounce in 1985 to 1,000 yuan ($270) by 1987, the prospectors know this is well below the prevailing international rate. Consequently, they hoard the gold or, more often, arrange for it to be smuggled out to Hong Kong for refining and barter against consumer goods.

The unofficial flow into Hong Kong has been accelerating. The Chinese authorities themselves concede that perhaps 10 tonnes a year escape their buying network. In reality, double that amount is filtering into Hong Kong's refineries. Threats of stricter controls and claims of 450 ounces confiscated in the first half of 1985 have done little to deter the private-enterprise spirit of China's rural gold diggers. The longer-term prospect, however, will depend less on the enthusiasm and energy of individuals and more on expanding the state-owned mining sector. The pattern of all gold rushes, ancient and modern, is that after a few years the individual gold diggers have gutted out the best and easiest part of the deposit. It then becomes a matter for more formal mining techniques. China will face this problem in the early 1990s. The authorities are quite aware of it. They have been talking with major Western mining companies for several years, picking their brains on new technology. But the Ministry of Metallurgical Industries has shied away from large-scale joint ventures with international mining houses. Instead, they have preferred only to utilise new technology, or to work with smaller companies whose experience is on their own modest scale.

New directions are already apparent. In Jiangxi province, in south-eastern China, which is billed as containing the country's richest untapped gold reserves, the provincial government is planning to open eleven state-operated mines before 1990. This is forecast to raise Jiangxi's output ten-fold. The Chinese have been looking closely, too, at the Australian and American gold mining scene to see what may be adapted from the host of small open-pit or underground operations now going forward around Kalgoorlie in Western Australia or in Nevada. The gold director at the People's Bank of China, which is paramount in determining policy on gold mining and gold sales, told me he had spent a couple of weeks in Kalgoorlie getting a first-hand look at that gold rush (see Chapter 5).

It was also no surprise to learn in Beijing that China has already installed at least five carbon-in-pulp treatment plants for low-grade ores; this is the very process on which Australia's success is founded. The Ministry of Metallurgical Industries has also set up a joint venture with Hunker Gold, a private Canadian company which runs a small mine in the Yukon, presumably to benefit from their know-how there. And the autonomous region of Xinjiang in north-western China has signed a joint venture with another small Canadian company, Galactic Resources. Galactic's task is to prospect the mountainous goldfields of Xinjiang where China's Geological Institute claims to have detected 4,000 orebodies. Galactic's chairman, Robert Friedland, responding to the Chinese challenge, enthusiastically suggested Xinjiang province might eventually offer 300-400 tonnes of gold a year.

That ambitious target was not in the sights of anyone else I talked to, but is a sign that China can become a serious long-term contender near the top of the gold producers' league. Unlike Australia, where the present boom may peak by 1990, I found China just getting into its stride; just starting to get to grips with tapping its mineral resources. Results do not come so swiftly in a centrally planned economy as in the free-wheeling environment of Australia or the United States (for all the diligence of China's rural prospectors). The government's decision, however, to switch significant resources into gold mining in the current five-year plan will form the foundation for further expansion. Output of 100 tonnes annually by 1990 is feasible, with continued growth thereafter for the rest of the century. Indeed, the full measure of China's gold resources will become apparent only in the 1990s. Yet already in this decade production has virtually tripled. If, more moderately, it doubled again between 1990 and 2000, then 200 tonnes a year might be achieved. That would make China the world's third-largest producer, after South Africa and the Soviet Union. Even if the Chinese people themselves devote more of their growing income to their traditional love of *kam*, the priority in a centrally planned economy is likely to be gold's value as foreign exchange. In short, China, like her Soviet neighbour, will become a regular contributor to world gold markets.

Part Two

THE ELECTRONIC MARKET PLACE

Chapter II

THE ELECTRONIC MARKET PLACE

The New York Commodity Exchange (Comex) would be a marvellous setting for a Broadway musical. All those traders in their colour-coded jackets - blue for aluminium, orange for copper and green for gold and silver - clustered around the respective 'pits' for each commodity, enmeshed in a web of stretched telephone lines to get their instructions for every continent. Huge electronic scoreboards behind flickering up the latest prices. A crescendo of noise fills the place. Cue for a spectacular production number called 'Limit Up' as a fresh frenzy of activity, sparked off by some rumour (quite possibly false), sends the gold price surging towards the $25 limit, up or down, over the previous day's price, that is permitted in a trading session.

Comex has added a show-business dimension to the world of gold, not least in the scale of turn-over. Over 8.4 million contracts were traded there in 1986, equivalent to 26,000 tonnes of gold (about a quarter of all gold ever mined) and worth approximately $300 billion. The flurry of activity often has little to do with what is really going on in gold. Ask a floor trader what motivates him and he may well reply, 'Greed'. Ask him what he knows about the supply-demand balance in gold or how gold jewellery sales are going this year and he will give you a blank stare. He wants turn-over, and it does not really matter if the price is going up or down. Comex is about 'buy' signals, 'sell' signals, or stop-loss orders, all dictated by charts and often ordained by computers. In this setting, the fact that Western gold mining production is soaring or that kilo bar shipments to, say, Singapore are down because smuggling to Indonesia is less, is irrelevant.

Comex is a magnet for every gold trader around the world during its opening hours from 8.20 a.m. to 2.30 p.m. In Geneva, if I call on a gold trader with many Arab clients, he pleads, 'Can you come before the market opens?' No need to ask which market. In Riyadh, the chief dealer at one of the main exchange houses explains he can only come to dinner after 9.30 p.m., when (on Saudi time) Comex closes. In Hong Kong, dinner ends, by comparison, abruptly at 9 p.m. as traders dash off to catch Comex opening. The world of gold waits on Comex and everyone is plugged into it.

What a contrast to the quiet meetings at N.M. Rothschilds in London at

10.30 a.m. and 3 p.m. each day, when the five members of the London gold market sit alone in an upstairs room beneath portraits of Nathan Mayer Rothschild
and European monarchs for whom the Rothschilds negotiated loans in the
nineteenth century, and by means of a single direct line to their own trading
rooms 'fix' the price. Little has changed there since the fixing started in September 1919 (except the price is now in dollars, not sterling) and most of the participants, Mocatta & Goldsmid, Samuel Montagu, Mase-Westpac, N.M.
Rothschild and Sharps Pixley, have an even longer track record. Mocatta &
Goldsmid started in 1671 and is still going strong. Only Mase-Westpac, who
have taken over the mantle of the ill-fated Johnson Matthey Bankers, who had to
be rescued from bankruptcy by the Bank of England in 1984 (due to problems on
their loan book, not gold), are newcomers.

Although Comex is the heavyweight and has added an extra dimension to
gold trading since 1975, the London fixing remains very much the bench mark
against which a great deal of real gold business is transacted; miners, central bankers, investors, fabricators, all know that if they use the fix, it is at a clearly posted
price, whereas on Comex the price shifts by the second. Moreover, the fix remains the one place to do a large volume at a single price. 'The world has an open
line to fixing,' says a London dealer, who has been plugged in for more than
twenty-five years. 'Anyone can participate for any amount through their broker. It
is *the* place for doing large amounts. Try to do one contract on Comex and the
price may move up against you.'

London's New Look

Yet the character of London is constantly changing. The immediate question
under discussion is how best to assimilate the other international banks and bullion houses, who have set up their own precious metal trading rooms in the City.
Their participation has broadened London's stature. The largest contingent is
American, including J. Aron (owned by Goldman Sachs), the Bank of Boston
(tied with Rhode Island National Hospital Trust), Drexel Burnham Lambert,
Morgan Guaranty, Philipp Brothers (part of Phibro-Salamon), Republic National
Bank of New York and Shearson Lehman Brothers (an American Express subsidiary). They have been joined by Bank of Nova Scotia from Toronto, and all
three main Swiss banks. The new Financial Services Act requires a more formal
market structure over which the Bank of England can keep a watchful eye, especially in the wake of the Johnson Matthey affair. 'The Bank of England doesn't
want to deal with half a dozen groups, we have to organise ourselves,' said Guy
Field of Morgan Guaranty. 'We must have an association with rules to govern the
behaviour of all traders. Up to now the role of the others, outside the five, was not
clear enough. It's a timely move, which will help London's stature as a recognised

international market place.' In the proposed reorganisation, the fix itself will be preserved with the famous five, but everyone will become members of the London Bullion Market Association operating under a Code of Conduct drawn up by the Bank of England.

This evolution has been essential to stay in tune with the changed nature of the gold game. Twenty years ago London was very much a physical market. The bullion packing rooms of the five houses dispatched gold bars of all weights and sizes to all corners of the world. South African production and most Russian sales went through London. The Swiss snatched that role away in a successful coup in 1968 when the London market was briefly closed. London had to find a different role. Building on the reputation of the fix, London has become a trading forum that acts both as a clearing house for the round-the-world, round-the-clock gold market, and the custodian of what is called 'loco London' gold. That is to say, gold which is traded and held in London on account of traders worldwide. The loco London facility quite simply avoids shifting gold all over the place. Exchanges like SIMEX (Singapore International Monetary Exchange) in Singapore quote a futures contract deliverable loco London; banks with golden passbook schemes for investors keep the necessary gold backing loco London; and market-makers themselves square positions loco London. Regional markets such as Hong Kong have a two-tier structure, with trading on their local exchanges complemented by a parallel market during their trading hours in loco London gold. There are arbitrage possibilities between a Hong Kong price in Hong Kong dollars per tael and a London price quoted in dollars. The successful image of loco London has encouraged the formation of a loco Tokyo quotation, too, which will become more widely accepted as gold stocks build up in Japan. But the cachet of loco London is not likely to be easily challenged.

The unique status of the Bank of England, not just as a recognised IMF gold depository, but as an even-handed agent in the gold market for many other central banks, also enhances London's position. The Bank, for instance, handled all the delicate transactions involved in the return to Iran of its gold, frozen in New York during the hostage crisis. The Bank's recognised diplomacy in such episodes means that many central banks still rely on it (often in conjunction with the Bank for International Settlements in Basel) to handle gold purchases or sales for them, and to store their gold. No country fears its assets being frozen in the Bank of England's vaults. This fact alone ensures London's future as a turntable of the international gold business.

Not that London is all a success story. The London Gold Futures market, unwisely starting with a sterling contract in a business that thinks in dollars, lasted three brief years. Its demise, however, taught a lesson that is worth consideration by other embryo gold markets. No natural basis existed in Britain for a

gold futures market. The British themselves have little interest in gold speculation, and are simply not accustomed to this kind of futures activity. By comparison, Americans have long loved to trade everything from pork bellies to soya
beans and orange juice on the New York and Chicago Exchanges. Gold added
just another chip in those casinos. The Chinese Gold and Silver Exchange Society in Hong Kong also has plenty of local punters, who love to speculate (especially on days when there is no horse racing at Happy Valley). The Tokyo Gold
Exchange is turning into an exciting forum as the Japanese are becoming increasingly attuned to gold trading. Turn-over in Tokyo of the 1-kilogramme contract
has shot up from less than 100,000 contracts in 1983 to over 1 million in 1986,
based almost entirely on local participation. Such activity provides the essential
liquidity to get a market off the ground. At the other end of the scale, both
SIMEX's 100-ounce contract deliverable *loco* London, and the Sydney Futures
Exchange link with Comex (essentially to extend Comex's trading hours) are ex-
periencing much heavier weather, because neither has a lively home team of
players.

Graduating from *Souk* to Global Market

All these markets, however, signal a radical change in the gold business. Once
upon a time it was physical business. Indeed, under the gold standard, gold coin
formed the basis of all money circulating within a country; you paid your bills in
Britain in sovereigns, in the US in Double Eagles, and in France in Napoleons.
The habit lingered in France for a generation after the gold standard, nourished
by war and devaluations, so that, until quite recently, the French had a great reputation as hoarders of kilo bars and Napoleons under the bed or in the cellar. Today
such a tradition lingers in many developing countries where physical gold, either
in coins, small bars or 22-carat jewellery, is a basic form of saving. Gold demand
in the *souks* - the markets - of North Africa and the Middle East is still an essential
ingredient of the market; and the 'mood of the *souks*' in judging when to buy or
sell gold can be as useful a signal as technical charts, as we shall discuss in later
chapters. But times change. Fifteen years ago the *souk* in Saudi Arabia received
the gold price, via Beirut, a day late by cable. Nowadays exchange houses and
banks in Jeddah and Riyadh all have their glowing Reuters money monitor screens, direct dial telephones and telexes. Communications, more than anything,
are revolutionising the gold business (as with other financial markets). *Loco* London gold, futures and options are natural developments as gold graduates from
souk to global market.

     That electronic market place increasingly dictates both the nature of the
gold business and the price. Kilo bars under the bed are giving way to a host of
gold-related financial instruments, which appeal to people who have never - and

probably never would - buy the metal itself. Their perception of gold also differs from the traditional one. They are buying gold to *make* money, not merely as a hedge against losing money. The thrust of the markets is to woo them in search of profit. Tempting ways of doing so are dangled before them.

'Financial service firms are becoming increasingly active as marketers of an expanding selection of retail investment products that are better suited to the needs and habits of the investing public,' says Jeffrey Nichols, president of American Precious Metal Advisors Inc. '... This growth in the gold market infrastructure is making investment easier for a larger and larger audience of mainstream investors.'[1] He contrasts this with the American scene a few years ago when 'gold availability was pretty much limited to small bars and bullion coins that could be purchased through a somewhat helter-skelter and motley assortment of coin shops, stamp dealers, and even gun shops, but not at reputable mainstream financial institutions.' The ordinary American investor was not interested. Today gold has been what Nichols called 'securitised' by paper-gold instruments that can be bought over the telephone, like stocks and bonds. The customer simply gets a piece of paper, a certificate that gold is held for him on metal account with a depository or a reputable bullion dealer. This opens new horizons. 'The mainstream investor, in contrast to the gold bug,' Nichols argues, 'prefers paper gold to the cash-and-carry business of physical gold. This mainstream investor is the market of the future.' The potential off-take in this new arena is enormous, not just in the United States, but throughout the world.

Price volatility increases. Once any punter, however modest, can buy or sell gold by picking up the telephone, he is more likely to run with the market, switching into or out of gold, not just daily but hourly. Specialist newsletters and technical analysts have a field day with players desperate for guidance. Professional investment groups programme their computers to signal crucial buy or sell points. Computer trading, in fact, is becoming a potent force - even a threat - that moves the market faster than more humble participants, denied such wizardry, can read the signs. 'The computer can be a deadly weapon to those who don't see it coming,' says John Powers, publisher of *Intermarket*, a Chicago magazine on global trading and risk management. Powers observes, moreover, that computers are often bearish. 'Computers are particularly feared,' he says, 'because they always appear to show up when the market is going down.' 'Computer-generated programs' sold the market down, newspapers say.[2] The triggering of 'stop-loss' selling by computers can create such an avalanche that people who thought they were protected by stop-loss at, say, $399, have found that order executed only at $380.

Small wonder then that gold seems less like 'money you can trust' and more like a casino chip. Gold's volatility is further exaggerated by the immense 'lever-

1 Jeffrey A. Nichols, 'Emerging Trends in Gold Investment', Gold 100 conference, Johannesburg, September 1986

2 *Intermarket*, vol.3, no.11, p.1, November 1986

age' created by futures contracts bought on a margin of only perhaps ten to twenty per cent, or options purchased for a premium that may be only five per cent of the actual striking price (for example, buying a six-month call on a strike price of $410 for a premium of $21). The attraction of options in particular is that the capital outlay is small compared to a physical transaction and that, unlike the futures market, the risk is limited to the premium paid.

Options Come of Age

The popularity of options is rising. Options on Comex, for instance, are up from a meagre 56,752 contracts in 1982 (the year they were first introduced) to 1,646,791 in 1986. And the cast of rival market-makers grows yearly. Originally, options were pioneered by Valeurs White Weld in Geneva and by Mocatta Metals Corporation in New York. Now there is scarcely a self-respecting bullion dealer who is not making a market in options. In Switzerland, Valeurs (re-named Credit Suisse First Boston Futures Trading) have been joined by all three big banks, Credit Suisse, Swiss Bank Corporation and Union Bank of Switzerland. In New York, take your choice from J. Aron, Drexel Burnham Lambert, Mocatta, Morgan Guaranty, Morgan Stanley, Philipp Brothers, Prudential Bache and Republic National Bank of New York; several of these houses also make markets through their London and Hong Kong offices. In London they compete with Rothschilds and Sharps Pixley. The European Options Exchange in Amsterdam has also lifted turn-over on its 10-ounce contract to nearly 250,000 annually.

Expanding membership of the options club is leading to a two-tier market. An individual market-maker cannot compete with the formal Comex option contract, so increasingly he has to offer 'over-the- counter' options for his customers, to whom he quotes special dates and special strike prices. 'We are orientated to do tailor-made options to meet a client's needs,' said a dealer at Credit Suisse First Boston. 'You want one for 2 p.m. on April 15 - we'll quote you. But that raises other problems. Our client may be the only one for that time and date, so who is on the other side?' The client increasingly demands good credit terms, too, and seeks out a market-maker who will sell him options without margin. This strengthens the hand of major American banks, who can give good credit facilities to certain clients, such as central banks, whom they know are a good risk. Central banks have not been slow to take advantage. 'Central banks and government investment agencies are now among the biggest speculators in options,' confided a New York market-maker. 'They have large reserves sitting idle and they want to get some revenue on them. So they sell calls above the market and earn the premium.' This tactic works well in a quiet market. In a volatile one, the central bank may risk getting called to deliver, but it is writing the option from a position of strength because it actually has the gold in its vaults. Such confidence enables central banks to operate on a grand scale. They may do 32,000 ounces (1 tonne) at a time, where the more usual option trades are 1,000-5,000 ounces.

Small players, however, can join in for a mere 100 ounces. In Europe, especially in Switzerland, many small banks and even local branch offices use the over-the-counter options market. A small Swiss bank, which specialises in accounts for farmers, is a regular writer of calls on their behalf, while excellent options business comes from the banks in Gstaad and St Moritz when the wealthy congregate there during the skiing season. 'The banks are all holders of metal', said a Swiss trader, 'so they seek to get a dividend on their gold stocks.'

Options are also beginning to attract gold miners, after considerable initial nervousness about the complexities of 'calls' and 'puts'. American and Canadian mining companies were the first to realise that they could guarantee themselves a minimum price for their production by selling puts (i.e. the right to sell the metal at an agreed price), but still reserve the opportunity of getting out if the price rose much higher than the striking price. Australian mining groups, who have tended to sell much of their output rather conservatively through the Gold Producers' Association auctions, are moving into options too. Looking ahead, market-makers believe participation by mines is the real growth area.

Many options players do not own gold. Speculators, including some well-known Swiss-based international commodity traders, write what are aptly called 'naked options' without owning an ounce of metal. Such vulnerability naturally adds to volatility if they have to cover when the market moves sharply against them. The leap in price from $290 to $340 almost overnight in March 1985, as gold swung back into the early stages of a new bull market, occurred partly because an international bank had built up an enormous 'naked' options position; so large, in fact, that at least one market-maker had already refused to risk any more exposure to them two months previously. Volatility comes not only from players who have sold calls and then are forced to cover. The price may be driven down by someone with a lot of puts which are due in two or three days at, say, a striking price of $420, when the market is at $460. He will try to push the price down to enable him to get out closer to the striking price.

Such manipulations spill over into the spot market and can, in the short term, have a major impact on the price. 'It becomes a snowball,' admits a Swiss dealer. The snowball effect may get larger, especially if gold miners write more options for price protection of their growing output. Rising liquidity, in turn, attracts a wider clientele on the other side. 'Options', added the dealer, 'have come of age as the third dimension of the gold market.'

The combined forces of futures, with the rising tide of options, are the real driving force for the gold price. As Paul Sarnoff, a wry, veteran of the markets, put it, 'The improbable truth when it comes to gold trading simply is that physical markets are now being led - and perhaps eclipsed - by gold futures and options markets. The tail is wagging the dog.' The reason, Sarnoff hastens to point out, is 'speculators'. Many punters, with a great deal of money at their disposal, have no interest in the physical gold market, but cannot resist a fling in the paper-gold market. The speculator buying a 100-ounce Comex futures contract does not want gold, he wants profits. He liquidates his position long before being placed in the embarrassing position of actually being obliged to take delivery. Scarcely one per cent of the gold traded on Comex is delivered and paid for.

A rising market, in particular, lures speculators like salmon leaping for the fisherman's fly. Monthly volume on Comex doubled, for instance, in September 1986 when gold moved back over $400 for the first time in three years, and soared again in April 1987 to 1.1 million contracts as the $400 barrier was breached. The injection of speculative money spurs the market to higher prices than could possibly be forecast just from the basic fundamentals or the daily doings on Comex of professional hedgers. 'After all, only twenty per cent of the action on Comex comes from hedgers, while eighty per cent comes from speculators who have to deposit only a pittance of the amount to buy contracts or sell short,' Sarnoff reminded me. 'Assume a good faith deposit of $3,000, then $60,000 would swing twenty Comex contracts representing 2,000 ounces of gold, which in the physical market would require a payment of $800,000.'

While speculators fuel a bull market, their desertion of a bear one inevitably drives the price into even deeper decline. I recall being in the office of a Geneva trader, who has many Middle East clients, one morning in 1983 when the London price fixed at just below a key chart point of $367. He excused himself, dashed into the trading room and started bailing out of gold. The price was soon below $350 and it took nearly three years to get back over that barrier. The speculators, meanwhile, were having more fun in those years with the dollar, financial futures, currency options, and the stock and bond markets. Deprived of their presence, the gold price limped along, often travelling in a narrow trading range of a few dollars for weeks or months at a time. Only the collapse of the dollar, nervousness that the boom in equities could not last forever, and anxiety about the American deficit and Third World debts brought speculators back in force in 1987. The gold price at once took on new life (while silver, which many had written off as a precious metal, also responded to the first transfusion of speculative money in several years).

A speculative run in gold certainly cheers up mining companies, and makes them even bolder in going ahead with marginal new mines. If they judge it right, they can lock in some nice profits by selling forward. But no miner can take decisions for long-term expansion based on the whim of the speculator. He has to look much more at the underlying true physical demand for gold. He must blow away the speculative froth, and see where the price may come to rest in the years when speculators are in other pastures. The miners got wonderful prices from 1978-81 but then had to survive several lean years. The immediate outlook in 1987 is also good, but between now and the year 2000 there will inevitably be many lean years when the price will have to rely on the 'bread-and-butter' demand. So, although the price will indeed rise to dizzy new heights on occasions, the long-haul outlook for a sustained rise in real terms is uncertain.

The Swiss Dimension

The arrival of the electronic marketplace must not obscure that fact. Futures and options offer professionals price protection and speculators gambling opportunities. But they are not about consumption. Someone, somewhere, ultimately has to pick up that gold, whether a central banker for reserves, an investor building a long-term portfolio, a goldsmith to make a wedding ring, or an industrial fabricator making gold-plated printed circuit boards for a computer, even a dentist for teeth. The gold price cannot ride forever on the back of paper-gold financial instruments.

The bright lights of Comex have not eliminated the need for physical gold markets, which stay firmly rooted in Europe and regional centres like Dubai, Hong Kong and Singapore, with Japan moving into its own separate league. New York traders often have a singularly narrow-minded view of the gold market; if Comex is quiet, no one is buying gold. On the contrary, I often remind them, there is a whole wide world of gold out there on other continents. Do you know how much was smuggled from Dubai to India last month? Or how much from Singapore to Indonesia? How are orders for gold chain in Vicenza and Arezzo in Italy? In those channels London and, above all, Zurich, make the running.

'The gold markets have moved to the United States and the very big players go to the big banks there - Morgan, Chase, Citibank,' conceded the head of precious metals for a leading Swiss bank. 'But the physical market is here. The Russians may use Comex to sell, but it's all ultimately unwound with EFPs (Exchange for Physicals) in Zurich.' The pre-eminence of the Swiss in actually delivering gold around the world in unequalled. They are the turntable. Switzerland imported 1,190 tonnes of gold in 1986, for example, compared with 461 tonnes for the London market, and 490 tonnes for the United States (where the figure was artificially high because of the routing of gold to Japan). Switzerland also exported 914 tonnes for distribution to regional markets, more than double the amount dispatched from London.

An interesting sidelight on these statistics is that they show a net build-up of stocks in Switzerland of 276 tonnes, the first increase for three years, which reflects a shift in favour of gold investment by Swiss money managers. The ebb and flow of gold into and out of Switzerland, in fact, is an excellent signal to market moods. During the bull market of the late 1970s, stocks built up substantially in Switzerland as banks increased exposure to ten or fifteen per cent of portfolios. This turned into net outflow in 1984 and 1985 as investors became disillusioned with gold and sold out, while the physical offtake in jewellery, especially in the Middle East and South-East Asia, picked up on low prices.

The positive build-up starting again in 1986 reveals not only a more favourable investment climate, but the dropping of the 6.2 per cent tax on physical gold sales in Switzerland itself. The tax, first imposed in 1980, had cut back sharply on over-the-counter sales of coins and bars in Switzerland (although it did not apply to metal account trading). Switzerland's image as a safe haven for gold investments was tarnished, and little tax was collected. So the government, being practical, repealed the tax, which was costing more to collect than the revenue gained.

Tax Troubles

The main beneficiary during the six years of the Swiss gold tax was Luxembourg, which blossomed as the one market in Europe which did not tax gold. Led by Banque Internationale and Kredietbank, a thriving market in gold coins and kilobars developed, particularly for German investors who faced an eleven per cent value added tax at home. Luxembourg became the forum in Europe for over-the-counter gold sales. Its future, however, is open to question with the Swiss tax removed.

Looking ahead, the threat of tax on gold could be a real thorn in market development. The imposition of value added tax on gold coin sales in both West Germany and Britain, for instance, reduced local demand. The hurdle of buying in offshore centres like Luxembourg or Switzerland does not deter the determined investor, but inevitably curbs the casual sale of a coin or small bar over the counter at home. The question mark now hangs over Japan. Although Prime Minister Nakasone had to shelve plans for a five per cent sales tax on gold during the Diet session of 1987 to break a deadlock over the passage of his budget, the long-term outlook is that tax will be introduced. 'It has only been postponed,' admitted a trader at Tanaka K.K., Japan's biggest precious metals house. Although the tax could be avoided by buying *loco* London gold, it would curtail physical sales. Many Japanese investors like to buy and hold their gold, often using it to hide 'black' earnings not declared for tax, and they are not yet familiar with the nuances of overseas markets. Their education, however, continues apace. Investors are being tuned in increasingly to the electronic market, which would enable them to bypass tax. Mitsui, a leading trading company, already has 5,000 members for its Gold Club, which takes telephone orders from 8 a.m. until 11 p.m. for as little as 10 grams to be held on metal account *loco* London. Their target is to get 100,000 members who could thus invest in gold exempt from any sales tax. The big securities houses, like Nomura, are also explaining to their vast network of clients the advantages of trading *loco* London gold.

So, the trend away from over-the-counter gold sales will continue in Japan and elsewhere. But the pace will be dictated both by communications and the need to convince people that 100 ounces of gold *loco* London really exists. In India that day is still remote; even trying to learn the gold price quickly is hard, and the Indian farmer still wants to put the profits of a good harvest into gold ornaments as an investment for bad seasons in future. Talking to dealers in Bombay's Javeri bazaar, the electronic gold market seems centuries away. They are more concerned about the latest smuggled consignment of ten tola bars - the chunky little gold bricks or 'biscuits' weighing 3.75 ounces - that every Indian prefers. India, that traditional sponge for precious metals, is a reminder of the way we were.

Change there will not come overnight. When I first visited India in the 1960s an economist at the Reserve Bank of India, who was working on new banking structure for the sub-continent, suggested the gold-hoarding habit would soon die out. People, he said, would use banks. Yet, twenty years on, the gold habit is as strong as ever.

The gold market, therefore, will proceed for the rest of this century on the dual track of its traditional physical markets and the brash new paper-gold ones. But the tide is running in the latter's favour. The ability to pick up a telephone almost anywhere in the world (except India) and direct dial to trade futures, options or *loco* London gold ensures that. But how will this dual market digest the new gold supplies? How, too, will 'bread-and-butter' demand provide a floor when the speculator is absent? In short, who are the real buyers?

Part Three

WHO ARE THE BUYERS?

Chapter 12

CENTRAL BANKS:
MONEY OR COMMODITY?

A fundamental shift in the character of the demand for gold has taken place over the last twenty years. Historically gold was a monetary metal. Its fixed price was a bench mark against which to judge commodities, and its prime function was as the means of exchange for both domestic and international trade. Gold was the cornerstone of the wealth of nations. That role ended finally in 1971 when the United States closed the Federal Reserve's gold 'window' and would no longer encash dollars handed in by other central banks for gold from its reserves at a fixed price. Yet long before that last plank in the gold standard was cut away, the role of gold had changed. The formal gold standard under which the ordinary citizen could redeem notes for gold coin at a fixed price had collapsed in the 1930s. The two-tier market for gold, created in 1968 to segregate official monetary transactions at a fixed price from the private market with a floating price, had ended a generation of gold at $35 an ounce. The whole thrust of public policy, especially on the part of the United States and the International Monetary Fund (IMF), was to divorce gold - John Maynard Keynes's 'barbarous relic' - from the monetary system.

In one sense, the divorce has worked. Central banks are no longer the major buyers of gold; they do not stand ready, as they once did, to buy gold from all comers at a fixed price. That shift is quite dramatic. Just consider the post-war years alone. Between 1948 and 1964, 7,897 tonnes of gold were purchased by central banks, representing 43.7 per cent of all the gold coming on the market in that period. By comparison, from 1965 to 1986 central banks were actually net *sellers* of over 2,000 tonnes into the private sector; they did not absorb one single ounce of newly mined gold. Obviously there were annual variations, but in only nine out of those twenty-two years were central banks net purchasers. The message is clear. The gold miner, for the first time in history, can no longer count on central banks or government institutions to buy his gold.

That is a radical reversal. In the strictest sense, it means that we must regard gold essentially as a commodity, not as a monetary metal. True, it is a commodity with a unique attraction for investors (who may include governments), but any mining company thinking ahead for the rest of this century and trying to project demand must focus principally on the application of gold as a commodity and private investment vehicle and not, as it might have done twenty years ago (indeed, as most did), as a monetary metal destined for the vaults of central banks. Gold is no longer the common denominator for currencies, and central banks do not absorb surplus production. Gold's role as a commodity is new, and is unprecedented in the long history of this precious metal. This is in direct contrast, for example, to platinum whose fortunes have always been founded on its industrial or jewellery applications, or even to silver, which has had to make its way as a commodity for well over 100 years, since most nations forsook the silver standard (in favour of gold) in the 1870s. Gold is still an infant as a commodity, still finding its feet.

The proposition that gold is a commodity is, of course, a highly contentious one. The more so because, although central banks no longer buy much gold, roughly one-third of all gold ever mined is sitting in government reserves. Official gold holdings amount to about 29,500 tonnes, with a further 6,100 tonnes held by the International Monetary Fund, the European Monetary Co-operation Fund and the Bank for International Settlements. Additionally, several countries, including such diverse places as Iraq and Singapore, do not declare their gold stocks, and a good deal is also held in the unpublished portfolios of various government investment institutions, principally in the Middle East (close to 1,000 tonnes, perhaps a trifle more, is in these secret treasuries). This gold represents not only a substantial share of international reserves today, but actually a bigger share than in 1971 when the formal $35 link was snapped. In that year, gold accounted for thirty-two per cent of international reserves; nine years later, with gold at $850, it represented an astonishing fifty-eight per cent of reserves. Over the whole period, it has averaged forty-five per cent of reserves. Remember, however, that this has been achieved with governments as *net sellers*; gold's status is all from the improvement of its price. Contrast this with Special Drawing Rights (SDRs), based upon a basket of five major currencies (US dollar, deutschmark, sterling, French franc and yen), which were conceived by the IMF as a new reserve asset. SDRs actually issued by the IMF account for scarcely 3 per cent of international reserves.

Government attitudes to gold in recent years have been that, for all their public pronouncements about wonderful new SDRs and demonetisation (as, indeed, it was officially by the amendment of IMF articles in 1978), privately they have largely husbanded their own stocks. The largest holdings remain with the major industrial nations, who have over eighty per cent of all official gold reserves. The high proportion of gold in the reserves of some countries is surprising; for the United States it is over seventy per cent, for Switzerland close to sixty-five per cent, for France, Italy and the Netherlands well over fifty per cent. Germany has around forty-five per cent, and Britain just over thirty per cent. The outsider is Japan with around twenty per cent, such a small slice that it prompts one of the key questions of the next few years: will Japan buy more to bring up its gold holdings? The other industrial shareholders, however, are of long standing, and none has any interest in increasing their stocks.

Official gold circles, in fact, have gone very quiet. As Dr Chris Stals, formerly deputy governor of the Reserve Bank of South Africa and now director general of finance for South Africa, who has been one of the closest observers of the gold scene, concedes, 'Today ... even the most ardent supporter of gold must admit that the events of the 1970s created a new international monetary order wherein the role of gold has been restricted to that of an important reserve asset, lying inactive and almost unused at the bottom of the pile of international reserves. The monetary role of gold is almost a forgotten issue.'[1]

The monetary debate has faded and central banks are not the positive force they once were in gold. Yet they must not be written off. Individually they use the gold market, at their own discretion, tailoring their activities to meet a wide variety of specific needs. Some sell local gold mine production, others buy to diversify reserves or even to prevent asset seizure for political acts (as with Iran or Libya) or non-payment of debts. Gold loans or swaps against foreign exchange may be entered into to tide a central bank over a balance of payments crisis. A few, like the Hungarian National Bank, are regular traders in the gold market, seeking to turn a profit, just as they might on foreign exchange dealings. Even the Bank of England runs a small position, ostensibly to cover the gold it requires to match sales of gold sovereigns, but also to test the pulse of the market.

Leasing gold reserves to the market has become fashionable. The gold earns the central bank anything from 0.5 to 1.5 per cent (usually paid in gold) on an otherwise unproductive asset, and gives the market additional liquidity with which to loan gold in turn to miners or fabricators (who find it cheaper than borrowing money). Several hundred tonnes will be leased into the market at any given moment. The Austrian National Bank is one of the keenest leasers, but it is a widespread custom, specially for small central banks anxious to get some return on their gold. Certainly, any analysis of the future of gold must consider the games central banks play, which can have considerable influence on the price from time to time.

The Active Players

The most open central bank activity is in gold-producing countries, with South Africa as the front-runner. The mining houses, by law, must sell their gold to the Reserve Bank of South Africa, which either puts the gold into reserves or, more usually, sells the production as it is delivered by the mines to about twenty banks

1 Dr Chris L. Stals, Keynote address, World Gold in 1986, London, June 1986

and bullion dealers in the key markets of London, Zurich, Frankfurt and New York. The Bank's role as the seller of 650-700 tonnes annually gives it special stature. It is in the market virtually every day as a seller, unlike the Soviet Union which is a regular trader but may not actually be a net seller for weeks or months at a time. The Reserve Bank's finesse, therefore, in judging the market's mood by being restrained when the price is weak and not selling too strongly into a rally, is crucial. Over the years the Bank has faced criticism from some international dealers, who feel it should do more to support a weak price - even to the extent of buying back. But the central bank has, until recently at least, resisted the temptation to become a trader, preferring the role of a flexible seller, and at best putting gold into reserves or engaging in swaps when the price is depressed. Signs of an important shift in policy became apparent in 1987, however, with the appointment of a new gold trading team at the Bank. For the first time, the Reserve Bank has started to engage in forward sales, futures and options. This has important implications for the market, if it enables South Africa to ride the crests and troughs more evenly, especially if she makes forward sales into speculative peaks, thus being able to hold off at other times. The new tactics are designed, of course, to gain South Africa the maximum price for her gold, but she is such a big player that they might help to smooth price volatility.

The Reserve Bank did not hesitate, for instance, to use the price peaks of late 1986 and early 1987 to unwind gold swaps in which it had engaged during the previous two years and to sell the gold at a good profit. The South African debt crisis in 1985 had forced her to engage in substantial gold swaps, involving close to 400 tonnes of gold, rather than sell the gold on the market with the inevitable depressing influence on the price. The swaps entitled the Reserve Bank to buy the gold back at some future date at a pre-determined price (in fact, spot plus interest) and either return it to reserves or sell it into a strong market (as happened in 1987). The unwinding of the swaps has important implications for the gold market. When the South African debt crisis was at its peak, bullion dealers were worried by the swap overhang. Suppose South Africa defaulted and the positions had to be liquidated? However, once the stronger price in 1987 enabled most of them to be redeemed, the overhang was eliminated. That is a bullish point for the market.

The use of gold by producing countries as a financial lifeline is on the increase. Brazil, Chile, Colombia, the Philippines, Venezuela and Zimbabwe are seeking to buy up part or all local output to stockpile or to market through their central banks. The advantage of this tactic is that they initially pay for the gold in local paper currencies, which are often depreciating fast in countries like Brazil, but swap or sell it for hard currency. Often central banks offer a substantial premium to get their hands on the local gold. Banco de la Republica in Colombia, for example, paid a thirty per cent premium over the market price, calculating in pesos at the black market exchange rate for several years prior to 1987. This margin was so attractive that gold was actually smuggled into Colombia through Panama for sale to the Bank, and the pesos then switched back to dollars profitably on the black market. Holding gold off the market in this way helps to provide a floor price, although in the long run the gold is not in very strong hands and it may be sold in any price surge or another debt crunch.

Brazil's experience illustrates how a debtor country can have a two-way influence on the price. The central bank first started buying local production in 1980, when significant alluvial gold discoveries rapidly pushed up output. The debt crisis of 1982, however, abruptly intervened. Brazil had to sell virtually all its reserve in the autumn of that year, but at least benefited from the high gold price that the crisis itself had sparked off. She got an average of $436 an ounce for her reserves. Their disposal also cooled the gold price. The central bank, however, rebuilt its stockpile, both from local output and on the international market during the low prices of the next three years. It was remarkably successful. Plucking up courage when gold fell below $300 in early 1985, for instance, the central bank bought nearly 25 tonnes and was one of the major reasons why the price held. In all, they had acquired over 100 tonnes of gold by September 1986 at an average price not much over $320. Then the resurgence of their debt crisis forced further gold sales, but at prices again over $400 and into a stronger market. Despite this reversal, the policy to buy up local production when possible will be maintained. The central bank and the government feel that, as a gold-producing nation, Brazil should have a good gold reserve. In short, they would like to be gold buyers when economic circumstances allow.

Banco de la Republica in Colombia has been purchasing domestic production much longer than Brazil, and initially built a stockpile of over 130 tonnes. It was then forced to sell much of it in 1984, but clawed back again by offering the high premium for the next two years. The Bank seems less concerned, however, to build a large reserve, preferring to get its hands on local gold and then sell it for foreign exchange when the price is right. At least, this means it will hold gold in moments of price weakness.

Venezuela has followed the example of its Latin American neighbours. The central bank established a gold-buying unit, Unidad de Oro, in 1986, to compete in the local market for newly mined gold and scrap. But, unlike Colombia, they have not paid a significant premium and purchases have been modest. Venezuela, incidentally, has by far the largest gold reserves in Latin America of 11.46 million ounces (356 tonnes), which include a good deal of valuable old gold coins, but the central bank has set its face against any gold sales from this reserve or the new purchases from local producers. The reason is that most of the reserve is in

Caracas; to move it abroad for sale would inevitably cause a political furore and a run on the bolivar.

The Philippines has also managed to stockpile its gold production, despite the political and economic upheavals. Domestic gold production must be sold to the central bank. All gold from primary mines is refined at the bank's own refinery in Quezon City, as is much of the alluvial gold which the bank's buying offices purchase in the countryside. Although the bank was forced to sell part of its reserve in the autumn of 1983 after the Aquino assassination, it has since managed to hold on to most of the domestic gold output each year. It acquired 24 tonnes in 1986, for instance. And with Jaime Ongpin, the former president of Benguet, the leading gold mining group, as minister of finance under President Aquino, the policy is set to continue. Ongpin is in no hurry to sell gold, and would probably want a price of $500 or more before he would consider selling a little of the reserve. Indeed, if the repayment of the Philippine debt can be placed on an acceptable basis (and Ongpin has proved a tough negotiator) and the economy can revive slightly, then no gold sales may be made at all.

Essentially, all these producing countries are reluctant sellers. They would prefer to husband their local output and dispose of it only in an emergency; the classic view, indeed, of gold as a lifebelt for dangerous times. Thus, looking ahead, we may expect these producers to provide a regular under-tow of buying when their economies allow it. The amounts may not be substantial, but it is possible for them to hold perhaps 50-75 tonnes off the market as they have done annually since 1983.

Avoiding Asset Seizure

Gold as a lifebelt also prompts buying by governments who have offended - or fear they may offend - the international community. They purchase the metal and take it home. Iran showed the way after her assets were frozen in the US hostage crisis in November 1979. Iraq and Libya have done the same thing. Their motives were political. More recently, Peru has done the same for economic reasons. She bought gold and took much of it back to Lima in late 1985 and early 1986 for fear that her assets might be frozen for non-payment of debt, after President Garcia announced he would only use fifteen per cent of export earning to service debts. The tactic was quite profitable; Peru bought the gold between $320 and $350, and sold some of it at over $420 some months later. Obviously such purchases are erratic; they are not a foundation stone of the gold market, but they show how central banks still feel a special relationship for gold in troubled days.

The real potential for official purchases of gold, however, has to come through diversification of reserves by countries which enjoy large surpluses but have little gold (which is, of course, what happened under the good old gold standard). A few years ago, the candidates were the oil producers. Besides Iran, Iraq and Libya, such OPEC members as Indonesia, Oman, Qatar, Saudi Arabia and the United Arab Emirates all bought. So did government investment institutions in the Gulf. They helped to soak up much of the gold sold by the US Treasury and the IMF, and provided a run of three years of positive official buying from 1980-83. The collapse of the oil price, however, in March 1983 removed them abruptly as the major buyers. So far, OPEC members, despite running down their reserves, have avoided selling gold (much of which was bought at very high prices) so they are neutral in the market.

Far East Potential

The balance of power has now shifted to the Pacific Basin. Japan, South Korea and Taiwan are making the running, aided by the little state of Brunei, which still enjoys a $20 billion investment portfolio from oil. Singapore could also come back, as its economy starts to recover from several years of slow growth. The vital question for the next decade is, can this region carry the gold market in the way that the Middle East, now a spent force, did from 1973 to 1983? The omens so far are quite auspicious, both in terms of government buying and in the private sector (as we shall consider in the following chapters). 'The Far East region as a whole has become ... a formidable consumer of gold,' notes Robert Sitt, managing director of Samuel Montagu in Hong Kong. 'It has really replaced the Middle East as the foremost market.'[2] He calculates that in 1985 the region absorbed almost forty per cent of all new gold supplies (or seventy-nine per cent of South African production), and in 1986 close to 700 tonnes or fifty per cent (and more than all South African output).

That performance was underpinned by Japan, which absorbed over 650 tonnes in 1986. The significance of the Japanese purchases was that they included 323 tonnes bought by the government to make the special 20-gram coin to commemorate the sixtieth anniversary of the Emperor Hirohito. Strictly speaking, this was not a central bank purchase, but it does represent the largest single acquisition by a government in many years. Moreover, the government succeeded in selling only 181 tonnes of the gold in the coins, so that it ended up with 142 tonnes of gold to spare (if not officially added to its reserves). Although the Japanese government made another coin issue in 1987 to use up more of the gold, it still has plenty in hand. It was also a very profitable exercise. The Japanese government got the first 223 tonnes of the gold at around $320-$330 an ounce, and the balance around $350. It sold the coins for 100,000 yen but they contained just under 40,000 yen in gold, so the government benefited from a premium of over 150 per cent. Indeed, they got 142 tonnes for stock free.

The success of this foray into gold could instil a more positive attitude

2 Robert Sitt, 'The Future of the Gold Price: An Asian Perspective', Gold 100 Conference, Johannesburg, September 1986

among both the Ministry of Finance and the Bank of Japan. For many years the Bank of Japan, wishing to keep in well with the US Treasury, deliberately avoided buying gold. But now gold forms scarcely twenty per cent of Japan's ever-growing reserves. This is much less than in all other major industrial countries, as we noted earlier. So will Japan make up the leeway? An aggressive buying campaign seems unlikely, but the pressures on Japan to improve her trade balance with the rest of the world may well encourage further gold purchases. The government went to considerable lengths in 1986 to window-dress their gold purchases to help their trade balance with the US. The gold all had to be delivered in New York, then was taken to Tokyo. The fact that the gold had mostly come from European dealers, because the order was far too big to be fulfilled from US gold supplies, did not show up in the improved US-Japan trade equation. Additional gold purchases would help the trade picture, and fulfil the dual task of allowing Japan to expand gold reserves.

Taiwan may also find diversification tempting. Her reserves have grown so fast that they are almost impossible to keep pace with. They cleared the $50 billion barrier in 1987 and kept surging. The Central Bank of China (as the Taiwanese central bank calls itself) keeps virtually all this nest-egg in dollars, and has only a modest stock of gold of perhaps 200-300 tonnes. A few purchases have been made, usually as a stock for special coin issues, but nothing to match the growth of reserves. The Bank for International Settlements's annual report, for example, noted 17 tonnes bought by Taiwan in 1985. The Ministry of Finance and the central bank in Taiwan take a very conservative view. They will not rush diversification, but the scale of their reserves has prompted international bankers to call on them with firm plans for asset management, which would include some holding in gold. Although no sudden shift is likely, a decision to put ten per cent of the reserve into gold would involve $5 billion, which would take up over 380 tonnes with gold at $400.

South Korea is also worth watching. Her reserves total over $3 billion but contain only 10 tonnes in gold (4 per cent). Like Japan and Taiwan, she has remained loyal to the dollar but its nose-dive since 1985 has put pressure on all of them to diversify. Already the government has agreed to a commemorative gold coin for the 1988 Olympics in Seoul. Bullion dealers see that as a foot in the door.

They have been treading an equally eager path to Bandar Seri Begawan, the capital of Brunei, to suggest to the Brunei Investment Agency, which runs a portfolio estimated at $20 billion derived from the little state's oil income, that some proportion in gold is prudent. The Agency has so far placed most of its assets in long-term government bonds and a mix of dollar, yen and sterling denominated equities. But, as the market mood turned in favour of precious metals in 1987, a shift out of bonds and perhaps some hedging of stock market profits in gold may be considered. (The Agency was highly amused early in 1986 when Japan first started discreet, but large, purchases of gold to find that many traders and analysts had decided it - or the Sultan of Brunei himself - was the big buyer. Telephones rang for days, but callers were politely, and correctly, told the Agency was not in the market.)

The financial muscle of these Far Eastern governments could have profound influence on the gold market during the next few years if one or more goes for gold. Not only would the actual amounts be substantial, but the psychological impact on the market would be considerable. Confirmation that Japan or Taiwan was a sustained buyer would give great confidence, because it is unlikely that they would be short-term traders; they would be acquiring for the long haul.

Spectacular surpluses in the Far East should not obscure a little discreet buying elsewhere, prompted particularly by the dollar's fall and unease about the debt situation. The central bank in Finland picked up 20 tonnes in 1985, judging correctly that the dollar had seen its best days. Nor must we neglect the role that the EEC's European Monetary Co-operation Fund assigns to gold. The participating European nations deposit twenty per cent of their gold with the Fund. In return, they receive currency units called ecus, which count as a reserve asset. The Fund holds nearly 2,700 tonnes of gold, and in calculating the issue of ecus values it at the full market price averaged over the preceding six months (unless the current day's price is lower, when it is adjusted downwards). Thus gold helps determine the quantity of ecus in the European system.

The European Fund's recognition of gold as a reserve asset shows that its unique status cannot be shrugged off. That perception was shared by Middle East government buying in the early 1980s. They put between five and eight per cent of their long-term investment portfolios in gold as a strategic stock, that is usually inviolable at the bottom of the reserve pile. The phrase 'strategic stock' crops up regularly, not just in the Middle East, but with other governments who like to have a little gold, often in their own vaults at home. Both Indonesia and Singapore, for instance, keep much of the gold they have acquired in their own central bank vaults. 'Gold is bedrock,' a London dealer told me when I started investigating the market in the 1960s; nothing has changed.

The Unofficial Gold Standard

In the strict sense, gold is not a monetary metal, but central banks seem very comfortable with it as a commodity (and it is the only one they hoard). The prime change, undoubtedly, is that much of the world's gold reserves are sterilised; there is no regular ebb and flow of gold between central banks in the industrialised world, reflecting their respective trade balances. (Imagine what would have happened to US gold reserves in the last few years of trade deficit if it had still been possible for other central banks to turn in dollars at the Federal Reserve for gold.) Yet, outside their orbit, a kind of unofficial gold standard is reasserting itself. In this chapter we have been talking about debtor countries producing and selling gold. Who buys it? A few years ago, it was the oil-rich nations of the Middle East, now it is Japan or Taiwan - with strong currencies and large surpluses. Should we say, 'The gold standard is dead, long live the gold standard'?

Do not forget either that an increasing number of governments have also authorised the minting of gold bullion coins (see Chapter 13). The way was paved by South Africa's krugerrand, but look who followed: Canada with the Maple Leaf, the United States with the Eagle, Australia with the Nugget, China with the Panda, Japan with the Hirohito, Belgium with the Ecu and Britain adding the Britannia to the sovereign that has been around since 1817. They may not be available at a fixed price, as the old gold standard demanded, but ordinary people can buy them.

James Turk, an investment adviser to a Middle East investment authority, puts an intriguing interpretation on it all. 'While the dollar and other national currencies are no longer redeemable into gold,' he argues, 'they are exchangeable for gold in the market place twenty-four hours a day. Therefore it would be correct to say that the dollar is on a flexible gold standard, and the rate of conversion is whatever the market price of gold is at that moment.'[3]

The catch in this view is that it ignores the strict discipline the original gold standard imposed on a country, which does not apply today. And it does not account, either, for the fact that central banks generally are not going out to buy gold regularly and in the quantities they once did. The prospect of greatly increased gold supplies for the rest of this century is not matched by central banks stepping out to pick it up just for the asking. Perhaps Australia's Reserve Bank might lend a hand to support its growing gold mining industry in moments of price weakness, but do not count on central banks in all gold-producing countries. The Bank of Canada, for instance, has a policy of reducing its gold stocks on any strong market rallies; it sold 12 tonnes in 1986 and kept on selling into 1987. Even the IMF, which sold 25 million ounces (777.6 tonnes) in the late 1970s to raise money for its trust fund for developing countries, is thinking seriously about selling some of the 3,217 tonnes it still holds to provide it with more liquidity to help the debt crisis.

Incidentally, proposals by various politicians and some newspapers (including *The Economist*) that central banks should dump gold on the market to drive down the price and bring South Africa to heel have little likelihood of being implemented. What they neglected to consider was not just the immense contraction in international reserves that a major fall in the gold price would imply, but that South Africa has, by and large, the lowest-cost mines. The mining industries

3 James Turk, 'Golden Sophisms', presented at An Evening of Financial Geology, New York, December 1986

of Australia, Canada, and the US, to say nothing of developing countries like Brazil, the Philippines, Ghana and Papua New Guinea, would be wiped out first. Was that really what the campaigners intended with such a proposal? This book has shown already how the balance of gold mining has shifted away from South Africa in the 1980s, to become much more widely dispersed. So the notion of central bank sales aimed at South Africa took little account of the realities of the world of gold today.

The reality of the central bank scene is that it remains a useful player in the gold market, but is certainly not shouldering the burden. The onus is on the growth of the private sector for jewellery, industry and, above all, investment. The test for gold to the year 2000 is not what central banks do, but how private investors perceive the metal. Do they still regard it as money?

Chapter 13

INVESTORS:
PORTFOLIO MANAGER
VERSUS SOUK

Building a bullish scenario for gold to the end of this century sometimes looks easy. The amount of money under management around the world is of such dimensions that only a trickle of it has to be diverted into gold for the price to take a quantum leap. The gold market, after all, operates on quite a modest scale: the amount of new metal reaching the market each year is between $23-28 billion, if gold is in the $400-500 an ounce price range. When the Japanese spent about $7 billion underpinning the gold market in 1986, it was hailed as a unique event (as indeed it was). Those sums, however, are petty cash compared with the glut of liquidity with which stock and bond markets are awash. World-wide funded pension assets total close to $3 trillion. The assets managed by America's 100 largest money managers alone were worth $1.5 trillion at the end of 1985, and had grown by $300 billion in that year alone. Or take a different comparison, from Julian Baring, a specialist in gold at James Capel, the London stockbrokers, who observed towards the end of 1986, 'If five per cent of the *profits* made in the US market alone since the beginning of 1985 were invested in the gold market that would amount to some $23 billion.' In short, a great deal of money is looking for a home.

Can gold provide at least part of that refuge? And what other factors come into play when it does? Selling gold is not a one-way street, as traders were reminded in the price surge of 1980 when immense quantities of gold - and silver - came flooding out of above-ground stocks in a matter of days to profit from the windfall. The physical gold market can go into reverse overnight, demanding that the investor absorbs not just conventional supplies, but dishoarding. Even in the modest run-up in price in September and October 1986, over $2 billion in gold was dishoarded in a few weeks from the Middle East and South-East Asia. The charts of this ebb and flow of gold, published here for the first time, give a unique insight into the 'real' demand for gold, and show how, as the price rises, it moves more and more into speculative terrain. If significant amounts of money shift into gold, that is not an immediate problem; the gold market, like any other, can be carried to great heights by the herd instinct.

During the 1970s, the era of hard assets, gold offered the investor an impeccable track record. The compound annual rate of return on gold in the ten years to 1980 was 31.6 per cent, compared to a mere 7.5 per cent for stocks, and matched only by Saudi Arabian Light oil. Unfortunately, the same cannot be said for the years 1980-85 when it declined on average 11.1 per cent. Gold has offered many attractive short-term trading opportunities in the eighties, but has not been a long-term asset. Even those who bought at below $300 early in 1985 have only really stayed all square because of the subsequent decline of the dollar, while the price in deutschmarks, Swiss francs and yen has languished. What has been missing has been a fresh catalyst to set the market alight, as oil did from 1973 to 1983. The oil boom created an era of inflation that pushed many investors into gold as a hedge, while also providing immense purchasing power to OPEC nations. Their central banks and government investment agencies, their private speculators and their ordinary grass roots gold buyers in the *souks* - the local markets of the Middle East - were the financial muscle in the gold market for a decade. What was lacking after oil collapsed in 1983 was a new driving force. That began to emerge only in 1987 in a combined cocktail of anxieties about the American deficit, Third World debt, the banking system, overheated stock markets, and South African upheavals.

The test is whether these anxieties will be crystallised into a new investment rush for gold. And how long would that last? It is a sobering thought that between 1968 and 1986 the so-called 'investor' was a net buyer of only 873 tonnes of gold, or 3.6 per cent of all supplies to the private sector in that period, according to Consolidated Gold Fields' annual surveys. Within that period, however, there were wide disparities. In the bull markets to 1974 and 1980 investors were substantial net buyers in the run-up, but then sold out. Since 1980, investors have been net sellers, except in 1983 and, marginally, in 1986.

The investor, however, is an elusive fellow. The term is often narrowly applied to people - or institutions - buying gold on metal account in Switzerland or *loco* London. The Swiss private banks, for example, who manage many discretionary accounts are clearly in this category. They stepped up their gold exposure in the late 1970s, sold out after 1980 and only really started to switch back in seriously during 1987. The real investment spectrum, however, is much wider. Broadly, it embraces three strands. First, the big investor, whom we have just mentioned, who may normally maintain four or five per cent of his portfolio in gold as an insurance policy, and increase that to ten or even fifteen per cent in a bull market. (One Geneva bank put its clients up to forty-five per cent in gold in 1980.) Second, there is the grass roots investor in Europe, America or Japan who tucks away gold coins or bars; the French once had the greatest reputation as hoarders, but are now being replaced by the Japanese. Third, there is what the Swiss banks used to call 'the traditional hoarding areas' in the *souks* of the Middle

East and the markets of South-East Asia, where gold coins, small bars and 22-carat 'investment' jewellery bought on a low mark-up remain the basic form of saving for the majority of the population.

The inter-play between these three strands of investment tells us a great deal about the future direction of gold. For all three never pull together. If the portfolio manager in Switzerland is building up his exposure to gold, then the *souk* is at best neutral and quite probably selling. And when shops in the *souk* are beseiged by buyers, and refineries cannot supply them with small bars fast enough, as in February 1985 with gold at under $300, the portfolio manager will have forsaken gold entirely and be playing currencies, equities or bonds. Charting 'the mood of the *souks*' over the years is actually a more reliable guide to gold than many technical charts, certainly in setting parameters for the gold price. If the local gold price in Cairo, Bombay, Jakarta, Bangkok, Hong Kong and Taipei is at a good premium to the London price, think about going long on gold, because that signals the floor. Conversely, if they are all at a discount, the international market is in speculative territory. This may not help the short-term trader thinking of the next minutes or hours, but is a signal to the more thoughtful investor.

The Mood of the *Souks*

The long-term outlook for gold, moreover, depends very much on 'the mood of the *souks*'. They are a loyal mainstay of the gold market for the long haul. Over the last twenty years, close to 8,000 tonnes of gold, or just over thirty per cent of all gold supplies to the market (excluding central banks), has been absorbed in a broad sweep of countries from Morocco along the North African coast to Egypt, Turkey, Saudi Arabia and the Gulf, the Indian sub-continent and South-East Asia. This off-take includes 18-carat gold jewellery exported from Italy, 22-carat locally fabricated jewellery, coins made in local factories (which make excellent replicas of everything from sovereigns to American Double Eagles), kilo bars and a wonderful range of small bars (tael bars for Hong Kong and Taiwan, baht bars for Bangkok and 'Fortunas' for the Middle East). They are sold on very low mark-ups over the gold price of the day, and represent the basic form of saving for millions of people in countries where banking systems, savings schemes and stock markets are not available (or are not trusted). Instead, hoarding in prosperous seasons and exchange of gold at marriages is still deeply interwoven into the social fabric.

Substantial amounts often change hands at marriage. In India five tolas (nearly 60 grams or just under 2 ounces) is still commonplace. In Morocco the bride will be adorned in a golden belt that may weigh anything from 200-300 grams to over 1 kilo. In Saudi Arabia many brides are decked with a *duru*, a lace-like embroidery of gold chain interwoven with small coins that often weighs over a kilo. At the modern factory of Saudi Gold outside Riyadh I saw one spectacular *duru* weighing 5 kilos which would drape the bride, like a cascade of light golden armour, from her neck almost to her knees. Such jewellery is purchased on a mark-up of only ten to twenty per cent over the gold price of the day. Consequently, a modest rise in the gold price produces a profit, and many women will trade in a few bangles, just as an investor takes his profit on a stock market peak. Equally, if they have cash and the price is low, they will buy more, gathering in the shops in the *souk* in the evenings for a social outing. They buy bracelets of 50 grams or 100 grams, often adorned with locally made sovereigns or half-sovereigns. Their husbands, meanwhile, will buy kilo bars when business is good (and the price low) and sell them at a profit if it rises or they need the liquidity. Specific local factors can push up gold demand even higher. In Turkey, for instance, coins and jewellery are the hedge against the constant devaluation of the lira. Over 55 tonnes of gold worth almost $650 million went into coins and jewellery in Turkey in 1986 alone.

Such habits die hard. When I first went to Beirut in 1967 to unravel the threads of the gold market web spun out from there, Turkey was already the main destination. Beirut, sadly, is no longer the centre of the web, but Turkey is still taking abundant gold. The pattern elsewhere in the Middle East, however, is changing simply because there is less purchasing power. In the heyday of oil, it was not only Saudi Arabia, Kuwait or Abu Dhabi that were great consumers but their neighbours like Jordan, Egypt and Yemen, through the earnings remitted by hundreds of thousands of migrant workers. Those days are over. Physical investment in gold bars in the Middle East dropped from a peak of 63 tonnes in 1982 to a mere 17 tonnes in 1986, while local fabrication of 'investment' jewellery (excluding Turkey) fell in the same period from 155 tonnes to 40 tonnes. No resurgence of gold-buying can be expected while oil remains in the doldrums. 'The money has evaporated,' said a banker in Jeddah, 'liquidity is very tight.' Indeed, any rise in the gold price has been seized as an opportunity for gold shops in the *souks*, many of whom have stocks of anything from 50-100 kilos of gold ornaments, to lighten up a little by selling 10 or 15 kilos for the melting pot. And what remains of the local fabrication industry is largely working on recycled scrap from this melted 'investment' jewellery. In Saudi Arabia, for example, half of all jewellery made in 1986 was from local scrap, and in Egypt all but 1 tonne of 27.3 tonnes going through the hall-marking offices came from scrap. When the price rose in the autumn of 1986, Saudi alone dishoarded over 75 tonnes of gold, almost all melted ornaments. 'You can write off the Middle East as investors for the next two or three years at least,' one of the main suppliers told me.

Less money is not the only reason for anticipating lower gold sales. The Islamic revolution continues its infiltration of many Moslem countries, subtly shifting attitudes. 'People are going back to the old school of Islam in which money and jewellery are less important in marriage,' a gold dealer in Jeddah reminded me. 'The couple go on Haj (pilgrimage) to Mecca instead. Before, every girl wanted some gold, now she believes in God and is not afraid at all.' Iran itself has vanished as a market for gold since the Shah's downfall, while the relentless spread of Shi'ite Moslem thought into the Gulf, Saudi and Egypt may well slowly lessen the gold habit. Coinciding with declining prosperity, this means that the Middle East is a spent force in the physical gold market.

What can replace it? Two areas remain strong: the Indian sub-continent, that traditional sponge for precious metals, and the Far East which, with help from Japan, has assumed the mantle of the Middle East as the prime consumer of gold. India's persistence as a gold buyer seems unshakeable. Indeed, going on to Bombay early in 1987, after two weeks of listening to depressing tales in the Middle East, was a revelation.

India: still a Social Symbol

The statistics posted monthly in neat white numbers on the black noticeboard behind the desk of the senior Bombay customs official looked impressive. Millions of rupees worth of watches, textiles, narcotics and gold had been seized each month. Over 2 tonnes of gold had been confiscated in 1986 alone, and 1987 started off with a sensational haul of 176 kilograms of gold (worth $2.2 million) hidden beneath a cargo of dates on a coastal dhow on the Bombay waterfront. The customs chief should have been pleased. Instead he confided sadly, 'In our heart of hearts we know that gold control has failed to shift the habits of our people. Gold is still part of our traditional social customs, it never changes.' Although India's customs and revenue intelligence officers caught over 2 tonnes of gold annually both in 1985 and 1986, over 100 tonnes (worth $1.2 billion) was successfully smuggled in. The seizure rate is scarcely two per cent.

India has forbidden the official import of gold since independence in 1947, except against special jewellery exports. And the government's Gold Control Administration has tried to wean the people away from their gold-buying habits. Initially, Gold Control sought to impose a switch from the popular 22-carat ornaments to 14-carat. No one paid much attention and that tactic was dropped. Now Gold Control monitors the activities of the sub-continent's 1,400 licensed gold dealers, trying to see they do not filter smuggled gold into the system. It also allocates a meagre 900 kilos of gold a year from India's own Hutti gold mine to a handful of authorised users in electronics, sari-making and Ayurvedic medicine. For the rest, the gold dealers and the army of 300,000 goldsmiths scattered through every Indian town and village get by with smuggled gold.

The Indian appetite for the metal ensures that the local price stays substantially above the international price and is not subject to its gyrations. The Indian price has advanced from 80 rupees per gram in 1963 (when Gold Control started) to 2,600 rupees per gram by 1987. In recent years, it has normally offered a healthy premium of $80-100 an ounce over the international price, which has made the sub-continent the favourite target for smugglers. In early 1987, for instance, while gold traded around $400 in London or New York, Bombay's price was around $510 at the black market exchange rate. The margin was irresistible. 'I see 40 tonnes needed for this marriage season in February, March, April and the first half of May; there must be gold to give,' said a trader in Dubai, the port at the southern end of the Arabian Gulf which has long been the main springboard for smuggling. Over the last twenty years, over 1,600 tonnes of gold has been imported into Dubai, all in the tola bars (3.75 troy ounces) that are favoured by India. Most of these 'biscuits', as they are dubbed, are spirited across the intervening water by high-powered dhows or with passengers and crew on aircraft. The 'biscuits' are dispatched slotted into the pockets of specially made light canvas jackets.

The remarkable fact about India is how little the demand has waned over the years: it remains one of the world's largest and most regular markets. While other regional markets of the Middle East or South-East Asia are very price-sensitive and stop taking gold on any sudden rise, India continues to soak up 80-100 tonnes a year. When I first investigated the gold market there twenty years ago, some Indians, especially government economists, felt the gold habit would be shaken off once banking and savings schemes reached into the rural areas where the farmers traditionally put the profits of a good harvest into gold ornaments. Nothing like that has happened. The stock market flourishes in Bombay now, but only about 150 stocks are seriously traded, and the players are mostly wealthy businessmen from the city. The rural areas stay faithful to gold.

'Gold is still the bastion, it is a social symbol,' explained a senior economist at the Reserve Bank of India. 'It is not just that it must be exchanged at marriage, it is the only way a villager in this country can get instant liquidity. He doesn't have a bank in his village, and he doesn't buy government bonds because they cannot be cashed instantly. Gold can.' Every Indian knows that if his crop fails or his family is sick he can raise cash in a moment from the village goldsmith. No one else can put rupees in his pocket without much frustration and paperwork. And the status of a family in its community is still judged by the gold that is exchanged as the bride's dowry in marriage. Higher gold prices have meant that the amount has come down in recent years from perhaps 10-15 tolas (3.75-5.62 troy ounces) to 3-5 tolas. But, given that there are nearly 10 million marriages in India each year, that alone calls for about 100 tonnes. Part of this is met by trading in old ornaments for new, but the underlying fresh demand is always there.

Social cachet and instant liquidity aside, gold is also a convenient anonymous haven for the 'black money' that Indians are always talking about. The black economy in India is variously supposed to account for thirty to fifty per cent of gross domestic product. Gold jewellery provides an ideal way of hiding and laundering undeclared income. Some Indian economists argue that all Gold Control rules should be swept aside in an effort to liberate some of this money and get it working productively in the economy. 'We need incentives to get people away from putting black money into gold,' said the Reserve Bank economist. The chances are slim. Initially, Prime Minister Rajiv Ghandi seemed ready to confront the gold issue. He even pencilled in on one memo, 'Why don't we scrap Gold Control?' Vested interests persuaded him otherwise. Yet the abolition of Gold Control within India and some tolerance of imports would not alter the ingrained social traditions. 'This country is wedded to gold,' says Shatilal Somawala, president of the Bombay Bullion Association. India, at least, is set firm as a cornerstone of tomorrow's gold market.

Counting on South-East Asia

The real test, however, is whether South-East Asia and Japan can sustain the leading role that the Middle East carried for a decade. The traditional affinity for gold in this region is with the Chinese, not just in Hong Kong, Singapore and Taiwan, but also in Indonesia, Malaysia and Thailand where, although a minority, they tend to dominate commercial life. Not only do they control the gold business everywhere, but their own communities are the greatest hoarders, because they often feel threatened as outsiders in a foreign land.

In Indonesia, Chinese businessmen in Jakarta or Surabaya, the main cities of Java, will buy gold at the first sign of communal strife. Before Indonesian elections, even the dealers in Japanese cars and electronic goods will run down their stocks and invest in gold in case of street violence. The story is similar in Thailand. 'People like to have a little gold. They feel safe,' said a Chinese dealer in his office above the bustle of Jarawad Road in Bangkok. Downstairs his shop was busy selling gold chains and the local baht bars (1 baht is 15.24 grams); the whole Bangkok market needs 2 tonnes monthly if the price is low.

In Taiwan gold investment has become a way of life, because of the permanent uncertainties about the 'Republic of China's' future against the might of the People's Republic of China just across the water on the mainland. Physical investment in jewellery and bars in Taiwan in the 1980s has absorbed nearly $5 billion in gold, more than any other country in the Far East, except Japan.

But will this high level be maintained in South-East Asia? The economy of Taiwan itself is strong, but Indonesia, Malaysia and Singapore have been through several tough years and their gold purchases have declined sharply. Investment holdings in gold passbooks run by banks in Singapore have tailed off, while in Indonesia the contraction of the economy due to weak oil prices has cut bar and investment jewellery sales. Indonesia, in fact, is a good example of the price sensitivity of 'investment' jewellery markets. Buying is strong between $300-400, but peters out quickly on any price rally. Dishoarding starts at over $400 and has been accelerated by the devaluation of the Indonesian rupiah, which led to the virtual doubling of the Jakarta gold price between early 1985 and 1987. While devaluation can be a recipe for buying gold (especially if it is anticipated), the high price after the event tempts profit-taking. And in a shrinking economy the high price inevitably erodes purchasing power. Thus Indonesia (an OPEC member) is, like the Middle East, less of a potent force in the gold market.

Hong Kong, which might seem an ideal centre for gold investment, given the uncertainties of its real status after 1997, also has an ambivalent attitude. While the debate on its future relationship with China raged in 1983, a great deal of gold was dishoarded locally as people shifted their assets elsewhere. Again, in 1986, they took advantage of the higher price to sell gold and get back into the local stock and property markets. Consequently during the 1980s net physical investment in gold in Hong Kong has been on a small scale; dishoarding in 1983 and 1986 has almost cancelled out buying in other years.

The price sensitivity in Hong Kong and elsewhere in South-East Asia, however, partly reflects the Chinese attitude to gold; they are in it for investment, rather than for long-term savings. That is to say, they will buy when the price is low and sell when it is high. This is classic 'mood of the *souks*' behaviour. They view gold as a two-way market. And because they tend to be contra-cyclical to the portfolio manager in Europe or America, they will be buying when he is engaged elsewhere.

This is the real significance of South-East Asia's gold markets in the long-term outlook for gold. The investors there provide the floor price. They will be picking it up at bargain rates when the New York view is that 'no one is buying gold'.

That point was driven home to me on a Saturday morning early in March 1985 in the gold market on Campbell Street in Penang. The gold price was hovering just under $290. I was going to visit one of Malaysia's leading dealers, Tham Yen Thim at Loy Hing Gold Merchant. The difficulty was to get into his shop, where the counters of gold ornaments were under seige from buyers. When I did battle through, he was almost too busy to talk. 'Gold is cheap,' he said. 'For the first time in years the housewives with a little money left over after shopping for food can buy a bracelet.' I knew then with absolute certainty that the bear market in gold was over. Three days later the gold price shot up in one afternoon to over $340.

The sheer appetite for gold in South-East Asia during those few weeks early in 1985, when people perceived it as 'cheap', was one of the most encouraging long-term signs for the gold price. In just three months, Hong Kong and Singapore alone imported nearly 150 tonnes of gold, equal almost to South African output in the period. The only hurdle was that European refineries simply could not keep up with orders for kilo bars, and eventually were quoting up to three months' delay on deliveries; a situation, incidentally, which suggested to an alert West German gold dealer that the best 'chart' to gold's direction might be the delay quoted by selected European refiners for kilo bar deliveries or, conversely, for processing dishoarded scrap. Three months' delay on kilo bars must signal the price near the floor, three months on processing scrap, the ceiling.

A similar guideline is achieved, in fact, by monitoring the imports and exports of gold through selected regional markets. Since the beginning of 1982 I have been tracking the monthly flows of gold through Dubai (mainly for India, but also a Middle East indicator), Hong Kong and Singapore. The results, shown in the chart below, are instructive.

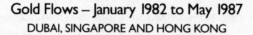

Gold Flows – January 1982 to May 1987
DUBAI, SINGAPORE AND HONG KONG

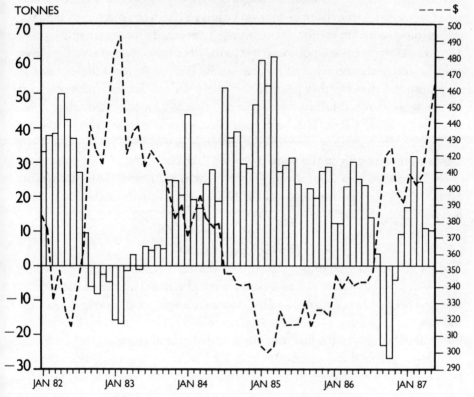

During the first half of 1982, when gold was drifting down from just under $400 towards $300, the combined monthly imports from Dubai, Hong Kong and Singapore rose by fifty per cent until June, when the price averaged only $314.97. But the moment the price suddenly shot over $400 (and briefly to $500) in the early autumn of 1982, demand in those centres not only evaporated, they actually started shipping back. So they switched overnight from taking gold at an annual rate of 500 tonnes to supplying it back at a rate of 60 tonnes. The dishoarding accelerated in January and February 1983, with gold at close to $500, to around 17 tonnes monthly (annual rate equivalent of 200 tonnes). This meant that markets which previously took most of South African production now took none, and were instead contributing gold at a rate close to normal Soviet sales. Clearly the price was too speculative to last. The correction duly came, but it was not until the price dipped below $400 in October 1983 that Dubai, Hong Kong and Singapore again started to import substantial amounts of gold. As the price then drifted down towards $300, the demand grew and grew. By January 1985, with gold averaging $302.82, these three centres imported 50 tonnes (more than all South African output). In the first three months of 1985, they absorbed fifty-six per cent of Western mine production. But the moment the price jumped to $340 in March 1985, the monthly imports fell away by over fifty per cent and never recovered. Once the price took off beyond $400 in the autumn of 1986 the dishoarding started in earnest again, averaging nearly 25 tonnes monthly.

These signposts are not much use to the floor trader on Comex, who lives only for the next moment, but they are useful both to the gold miner and the long-term investor. For they point to the 'safety net' for the gold price provided by the demand in South-East Asia, India and also, to a more limited extent these days, in the Middle East. This 'safety net' checks a falling price when people see gold as 'cheap'. Certainly the lesson of the *souks* is that they get the long-term trends right. Not only in the years covered by the chart, but in the previous decade, their judgement was correct. They bought in 1970 at $45, sold out close to $200 in 1974, started in again at $110 in 1976, and sold out in 1980. That pattern is repeated constantly.

The Golding of Japan

The future task for gold miners, so busy producing ever more gold, is to address themselves to that 'safety net' to see how it may be raised a trifle all the time to keep the floor price edging up. In 1970, $35 was cheap; in 1985, so was $300. Can the 'safety net' be hitched up to the $400 notch and held there? Much depends on two other elements in the investment scene: Japan, and coins.

Japan stunned the world of gold in 1986 by absorbing 677 tonnes of gold bars, coin and jewellery worth over $8.0 billion; that was more than all South African production, and forty per cent of combined Western mine output and Communist sales. Private Japanese investors alone picked up 376 tonnes of the gold, in coins and bars, while the balance went into jewellery, industry and government stocks. The purchases put Japan in a league of her own, underpinning the gold price, and providing the springboard for the leap through the $400 barrier.

Was it a unique year or merely a foretaste of things to come? Can or, rather, will the Japanese investor shoulder the burden imposed on the market by increasing gold supplies? He has the resources to do so. The Japanese have one of the highest savings rates in the world, measuring seventeen per cent of disposable income in recent years. Private investors (as distinct from institutional) alone had 550 trillion yen ($3.4 trillion) at their disposal by the end of 1986, but only the smallest slice - 0.35% - was in gold. The potential, therefore, is enormous.

Japan's exceptional consumption in 1986 was exaggerated, of course, by the special 20-gram coin with a face value of 100,000 yen issued by the government to mark the 60th anniversary of Emperor Hirohito's reign. Just over 9 million of the coins, comprising 181 tonnes, were actually sold; a remarkable achievement, since the premium was 1½ times the gold content. However, since the coins are interchangeable with 100,000 yen banknotes, there is actually no downside risk. The Hirohito issue was, however, unique in quantity and, although the Ministry of Finance made a second modest distribution of 1 million Hirohito coins (including 300,000 proof quality) in 1987, nothing on such a grand scale is planned in future.

The best way to view Japan is to ignore 1986 entirely and look at the progression of gold investment in other recent years as a clue to future intentions. The essential point to remember is that the Japanese are relative newcomers to gold. Their gold market has only been fully liberalised since April 1982, when banks and securities houses were finally allowed to start gold trading. Although gold had been available to private buyers for nearly ten years before that through the leading precious metal houses, Tanaka, Tokuriki and Ishifuku and such trading companies as Sumitomo, it had not reached a wide investment audience. Investors have been cultivated only relatively recently. The initial response has been encouraging. They have absorbed over 100 tonnes annually since 1984.

However, once the yen began to assume its ascendancy over the dollar, causing yen gold prices in Japan to tumble, the real surge started. As the gold price came down through a series of price levels, 3,000 yen a gram, then 2,500 yen, then 2,000 yen, so demand picked up. The Japanese also regarded gold as 'cheap', just as people in South-East Asia had done at under $300. The enthusiasm for bargain-basement buying is a pointer to the present Japanese attitude to gold; they are in the early days of building up private stocks of gold for their long-term savings. Where the Chinese buying in Hong Kong mostly have their nest-egg and trade gold off the top as an investment vehicle, the Japanese are still acquiring the basic nest-egg. Investors have already squirrelled away nearly 800 tonnes, but that is petty cash in such a wealthy country. Serious money has gone into the stock market, real estate, even into buying golf club memberships (which are traded in a secondary market as negotiable certificates, often commanding prices over $100,000 for the classier golf clubs). Gold is only now starting to compete.

This makes Japan one of the most bullish factors on the gold horizon for the rest of this century. The Japanese people are suddenly aware of their wealth, and are becoming fascinated with the concept of investment diversification to spread risks. 'Personal financial management has become a daily topic of conversation among office workers, housewives, and even students,' explained Itsuo Toshima, formerly Swiss Bank Corporation's Tokyo gold trader and now Far East investment manager of the World Gold Council. 'The word *zaiteku*, which literally means financial technology, has become a household word. The idea is to apply to personal money management the same spirit and degree of expertise which created the world-famous Japanese industrial technology.'[1]

That initiative is directing more and more people to consider gold. One significant factor about the Hirohito coin was that it put a gold coin into the hands of 9 million of Japan's 30 million families, most of whom had never bought gold before. The incentive to follow up may be strong. Interest rates in Japan are very low: only 3.76 per cent on one-year time deposits, and a mere 0.26 per cent on ordinary bank deposits. Consequently, bank deposits, where Japanese investors keep fifty per cent of their assets, have little attraction. Banks may have even less appeal if the tax authorities go ahead with plans to cancel tax exemption on the interest received on deposits of up to 3 million yen (about $20,000). Many individuals have got round any tax liability by opening several bank accounts, often in fictitious names, and keeping the balance of each under 3 million yen. The ending of tax exemption (designed originally to encourage small savers) has already touched off a flow of funds away from bank deposits into other forms of investment.

The sparkle of the Tokyo stock market initially captured much of the money, but the question of what would happen when that boom ended was already on many people's minds when I was in Tokyo in the spring of 1987. 'Today many private investors have made a good profit from the stock market,' Yoshio Sekine, managing director of the precious metal firm Tokuriki, told me. 'But if the stock market declines, then do they switch into gold? If they do, they could be very big buyers.' His forecast was correct. Within a few months, Tokyo dealers were witnessing a new phenomenon. 'It's a totally new breed of gold investors,' Itsuo Toshima informed me, 'who feel uneasy about the exaggerated stock prices and start shifting their funds into other investment vehicles, including gold.' Individual investors, buying through securities companies, have picked up as much as 200-300 kilos at a time. Japan's imports, which had languished, suddenly improved again to 25-30 tonnes monthly by mid-1987. Tokyo traders also observed buying on behalf of the treasury departments of up-and-coming young Japanese companies which do not have the large hidden reserves, mostly in property, enjoyed by older firms. 'These new companies lack tangible assets, so they buy gold,' a trader confided. 'I know of 5 tonnes bought through securities houses.'

The securities companies are working hard to exploit this avenue. Daiwa Securities has already formed a subsidiary, Daiwa Precious Metals, solely to explore institutional investment in gold. But there are restrictions on financial institutions, and most money managers do not want to hold gold as insurance; they prefer short-term profit in bonds and equities. So the private investor remains the prime target. The trading company Mitsui is already wooing them through its Gold Club to buy and sell gold *loco* London, as we saw in chapter 11.

Yet a surprising amount of the physical gold market in Japan is still through twenty to thirty retail outlets, controlled by Tanaka, Tokuriki, Ishifuku and Sumitomo. The main turn-over in all of them is in 500-gram and 1-kilo bars, because they are tax free. Gold coins attract a luxury tax if they cost more than 37,000 yen (which rules out anything over half an ounce). 'We've got four retail shops in Tokyo and Osaka,' Akio Imamura of Sumitomo told me, 'and the number of customers increases every year. They buy anything from 100 grams to 40 or 50 kilos in our shops, and almost all of it is for the long term. Maybe twenty per cent is short-term speculation.'

Japan's ultimate gold shop is Yamazaki, part of the Tanaka empire, on Ginza in Tokyo. Pop music and a big TV screen featuring a fashion show of gold jewellery greet the customer. The day's prices for kilo bars, Maple Leafs, Eagles and Nuggets, and 22-carat gold jewellery flash up on an electronic screen, while a huge map of the world spotlights the latest gold price in key international markets. Yamazaki has four floors devoted to gold. The ground floor is given over entirely to chain sold by weight. 'We offer 500 kinds of chain,' said Tadahiko Fukami, one of Tanaka's senior managing directors, proudly guiding me round. 'In all we have over 20,000 kinds of jewellery, including rings, bracelets and pendants. At the peak times we have 2,000 customers a day.' The basement is given over to rings and pendants, and the second floor to European jewellery. The price of each item is quoted in gold and in yen, so that customers can, if they wish, pay in gold. 'At this shop, you can use gold as money,' said Mr Fukami. A buy-back price is also quoted, so that jewellery can be traded in.

The serious gold investor, however, takes the elevator to the third floor for a rich display of coins and bars. The latest prices for 9999 gold in Tokyo, Hong

Kong, Zurich, New York and London flicker up on an electronic scoreboard. 'Today's selling price is 2,040 yen a gram,' said Mr Fukami, 'and our buy-back is 1,940 yen.' The price meant Yamazaki was relatively quiet; 2,000 yen is a psychological barrier. Investors were hoping it would slip below before buying again. Yamazaki can afford to wait; the store's turn-over in 1986 was an astonishing 40 billion yen (about $250 million).

The store's success shows how serious the Japanese are in their love affair with gold. I have never encountered such a concerted drive to woo the gold buyer, whether to jewellery, coins or bars. Nothing of the scale of Yamazaki is to be found in Europe or America. This highlights, too, the extent to which Japan is still very much a physical gold market. Although volume on the Tokyo Gold Futures Exchange has been rising fast, the players there are principally professionals and a small band of speculators. Most Japanese, however, are nervous of such paper gold markets, and with some reason. In a major scandal in 1985 a company calling itself Toyota Trading (having absolutely no connection with Toyota Motors) issued gold certificates, but neglected to back them with physical gold. When called on to deliver, they could not. Customers lost $1.2 billion, equivalent to over 100 tonnes of gold. The episode ended with the president of the company being stabbed to death by a distraught investor in front of journalists and television cameras. Not surprisingly, this turned the Japanese off paper gold. They prefer gold bars.

So Japanese gold demand will continue to be satisfied largely by physical delivery. I asked Jun-ichiro Tanaka, the president of Tanaka Kikinjoku Kogyo, Japan's foremost precious metal traders (who celebrated their centenary in 1985), what was his outlook for Japanese buying for the rest of this century. He took the question seriously, and when I called on him in Tokyo, there was a detailed paper already prepared. He stressed, first, that 'investment in gold by individuals will still be concentrated upon the spot transaction'. Then Mr Tanaka went on to explain that the present amount of gold held by Japanese was 'too small' when compared with private holdings in France or the United States; it was not yet 1,000 tonnes. 'The proportion of gold stock in the total monetary assets held by individuals is predicted to reach 3,000 tonnes,' he concluded. That implies another 2,000 tonnes or more still to be acquired. Investment adviser Itsuo Toshima was even more bullish. He based his assumption on the fact that gold forms only 0.35 per cent of investment assets in Japan, compared to 1.59 per cent in West Germany. Since in other respects in portfolio breakdowns the two countries are similar, he estimates the Japanese must buy another 3,000 tonnes at 1987 price levels over the next fifteen to twenty years to catch up. That amounts to an annual average of 150-200 tonnes.

Whatever the precise amounts, the message I picked up in Tokyo is that

Japan's involvement with gold is still in the formative stages, and it can be perhaps the single most important force in investment gold for the rest of this century. The clue is that word *zaiteku*, financial technology. If Japanese investors can apply the diligence that has given them such leadership in electronics, cameras and cars to managing their own money, watch out for gold. It is like a snowball at the top of a hill, starting its roll, gathering momentum and growing larger and larger every second. That is what is happening to gold in Japan.

The Small Product Revolution

The most radical initiative in bringing gold to a broad spectrum of investors has been the creation of the legal tender bullion coin, sold on a small premium over the day's spot price. The bullion coin has brought gold within the means of the man or woman in the street. The pioneer, of course, was South Africa's Kruger-rand, which was the only serious contender for a decade or more until it fell foul of political sanctions. But its success spawned Australia's Nugget, Canada's Maple Leaf and the United States's Eagle as major competitors, and put a gleam in the eye of many other countries ranging from Belgium with its Ecu, to China with the Panda and the United Kingdom with the Britannia. Japan, of course, joined in with the special Hirohito coin in 1986, which was not a bullion coin but a mighty consumer of gold.

The significance of these newcomers should not be under-estimated. As a Miami coin dealer put it to me shortly after the Eagle launch in the autumn of 1986, 'The US government is basically saying, It's okay to buy gold.' Which was quite a reversal for a nation that banned private gold holding by its citizens from 1934 to 1975. Australia's Nugget received a similar blessing from Prime Minister Bob Hawke at its launch in April 1987 when he noted, 'Australia is now the world's fastest-growing gold producer, so it makes good sense that we should be one of the world's *top* producers of bullion coins the coins will be legal tender in Australia so buyers can be confident of their quality and standard.'

The comeback of gold coins has opened up a major sector of consumption. Until the Krugerrand took off in 1974, fabrication into official coins was, at best, modest. Coins had become more of a collector's than an investor's item. No more than 50-60 tonnes were minted annually and limited mainly to Austrian ducats, Mexican centenarios and British sovereigns, which all commanded comfortable premiums. The Kruger, with a premium to distributors of only three per cent over the spot price, introduced a different dimension. In the thirteen years from 1974 to 1986, gold coin fabrication consumed over 2,790 tonnes of gold, or sixteen per cent of all gold supplies to the private sector. Coin demand usually now exceeds combined industrial and dental use. Although Japan's Hirohito coin lifted fabrication to a record 316 tonnes in 1986, the track record for the regular bullion coins alone is impressive; they have accounted for almost fifteen per cent.

The demise of the Krugerrand, which has not been minted since 1985 when imports into the United States and Europe were halted, has not dented the market. Indeed, it seems to have taken on a new lease of life. Even ignoring the Hirohito, coin fabrication in 1986 was higher than in the previous two years. Although the spate of new coins may appear merely to be fighting for a slice of the same cake, in practice they are actually broadening the market by constantly tapping new investors who, for various reasons, had not bought coins before. As long as the market was dominated by the Krugerrand, there were many investors in the United States, especially in the New York, New England and Chicago areas, who would not touch the coins for political reasons. Many American banks, too, would not handle Krugers. But once Canada launched the Maple Leaf in 1979, it immediately found favour with many first-time American investors. Since 1986, the US Eagle has widened the spectrum further. 'New people who never touched gold before want the Eagle,' said Luis Vigdor of Manfra, Tordella and Brookes, one of New York's top coin dealers. 'The yuppies like it.'

Goldcorp Australia, which is marketing the Nugget, seeks a similar fresh clientele in Australia, where bullion coins have previously made no impact. 'We've got all 1,300 branches of Westpac Banking distributing the coin,' Don Mackay-Coghill, Goldcorp's managing director, told me. The Nugget is also being promoted heavily as a symbol of 'good fortune' in Taiwan and Singapore, where gold coins have not yet established a firm foothold.

Inevitably coin sales depend on the investment climate. Investors see them as a hedge against inflation or currency devaluation, and a refuge from other political or economic anxieties. The heyday of the bullion coin was in the inflationary days of the late 1970s. But a hard core of demand of 100-150 tonnes has persisted year in, year out. The new coins have helped maintain momentum because many investors buy the first year of issue, hoping they will eventually command a special premium. The US Eagle notched up 55 tonnes of sales in its first three months in 1986, as Americans scrambled to get their hands on the first US legal tender coin in over fifty years. Goldcorp Australia sold 155,000 ounces (4.8 tonnes) of Nuggets the very first day. Novelty may later breed loyalty. China's Panda coins have carved out a dedicated group of collectors, who buy each annual issue because the design changes (unlike the Krugerrand, Maple Leaf, Nugget or Eagle, whose design is permanent).

Much agony goes into coin design, with each promoter seeking some distinctive appeal. Canada was able to tout the Maple Leaf as the first 'pure gold' 9999 coin against the Krugerrand, which was only 916 fine and has a rather coppery tint from the associated alloy. The Nugget is lauded for its crisp image of the 'Welcome Stranger' nugget standing out against a frosted background. The four

Nugget coins - 1 ounce, ½ ounce, ¼ ounce and 1/10 ounce - also have special appeal to collectors, because each features a different famous Australian nugget (the 'Welcome Stranger' on the 1 ounce was the largest ever found, weighing in at 2,284 ounces).

The main benefit of a range of sizes, which all the bullion coins now offer, is not only that the small ones can be afforded even by the impecunious investor (or can be used in cufflinks or other jewellery), but that they fall just below the level for luxury sales in Japan. This cuts in at 37,000 yen, and so far has virtually eliminated sales of any 1-ounce bullion coins. The coin promoters eye Japan assiduously, and have campaigned hard to get coins tax-exempt to compete on equal terms with bars, which are all tax-free. So far, their lobbying has come to nothing, but a market approaching 15 tonnes a year has been carved out for the ½ ounce and smaller coins, with the Maple Leaf well in the lead. The situation in Japan may change. If the luxury tax is replaced, as many forecast, by a general sales tax of perhaps three or five per cent, then coins and bars would be on equal footing, and coins might win a larger share. In the aftermath of the Hirohito, they may tap a broader vein of investors anyway. So far, investment in Japan has focussed on wealthier people buying 500-gram and 1-kilo bars. Looking ahead, the test for the bullion coin market is going to be three-fold. Will it be more heavily taxed? Can a better network of distribution be established? And can coins become more widely used as a part of the ordinary investors' portfolio?.

Tax inhibits coin sales. The direct market for Krugerrands in Britain and West Germany was largely killed by the imposition of value added tax, which was avoided for several years by the 'legal tender' status. The French market, too, was snuffed out by tax and the ending of anonymity of gold sales. A successful coin market has been established in Luxembourg, where Banque Internationale and Kredietbank, in particular, have built up an international 'offshore' clientele. The Swiss market is also open since the abolition of tax in October 1986, but there is really no substitute for over-the-counter sales. The small investor in, say, Britain or France is not going to buy 'offshore' and indeed would probably not know how to. Neither is he willing to pay fifteen per cent or more sales tax at home. The European market, therefore, is in something of a straitjacket. Moreover, attitudes to physical gold investment in Europe, which were originally shaped by two world wars ravaging through France and West Germany, have changed.

The younger generation in France pays little heed to gold coin investment, although their parents were brought up to hoard Napoleons. Twenty years ago, in the days of General de Gaulle praising gold as 'the unalterable fiduciary value *par excellence*', the French were indeed an important factor in the gold market. The daily price fix beneath the Bourse in Paris was a major event. Today it passes almost unnoticed in the wide world of gold. The French may still have a private hoard of over 5,000 tonnes, but it is not growing. They are, of course, hedged in by exchange controls which also officially prohibit gold coin imports. If controls were lifted, then some comeback might be achieved, but the basic motives of fear of invasion and constant devaluation, which nourished the habit, have gone. De Gaulle gave France the confidence and stability that eliminated the need for gold.

The target for the bullion coin promoters has to be the United States, Japan (if the luxury tax goes) and perhaps South-East Asia which is still more attuned to local gold bars. Two things now make the task easier in the United States. First, the motley assortment of coin shops through which gold coins were initially sold after liberalisation of the gold market in 1975 is now being replaced by a more solid distribution network. All the major brokerage houses like Merrill Lynch, Dean Witter, Bache and E.F. Hutton have coin programmes for gold on metal accounts, or delivery to depositories in Delaware. Better banking networks are emerging. Howard Glicken of Metalbanc in Miami told me how he was hoping to line up a thousand local banks and savings and loan associations in such diverse places as Topeka, Kansas, Charlotte, North Carolina and Caspar, Wyoming, whom he could service with Eagle coins. The Eagle has also won additional respectability because, since the beginning of 1987, it has been permitted as a tax-deductible item for Individual Retirement Accounts (IRAs). 'The Eagle is the only gold that is allowed in an IRA,' said a New York dealer, 'but it means the US government is standing behind gold as an investment.'

The significance of this development is that for the first time gold can form part of the regular investment portfolio of Americans. Its advocates so far have tended to be a rather narrow range of dedicated 'gold bugs', located in the southern and south-western states. Although institutional money managers in general are still reluctant to touch gold - they are too concerned with short-term performance - a subtle shift is detectable among the wealthy private investors and their advisers. 'We get calls now from people running family trusts', confided a New York banker, 'to put half a million dollars or a million dollars into coins.'

Portfolio investment in gold coin is nothing new. In Switzerland many of the smaller banks, which may have five to ten per cent of portfolio in gold, will include a slice of that in coin. The advantage of coin over, say, a 1-kilo bar is it gives flexibility to sell five or ten 1-ounce coins at a time into a price rally, rather than have to sell a whole kilo at a single price. Coin premiums can also give extra benefit. Several alert Swiss banks noted in the summer of 1984 that the premium on the Swiss vreneli coins, which are in limited supply and normally command a premium of twenty-five to fifty per cent, had slipped to a mere four per cent - no more than bullion coins. So they sold Krugerrands from portfolios, and switched into vreneli. A shrewd move, because the vreneli premium later went back up over twenty per cent. Conversely, Krugerrands can still be attractive in a portfolio because, although they are no longer minted, the pool of over 40 million in circulation trades at a premium of less than one per cent. Krugers are thus a way of getting into gold coin at virtually the spot price, with little downside premium risk because the worst that can happen is that they go to 'melt' price, which is about a half a per cent under spot.

The importance of bullion coins in creating a new market for gold has been accompanied by a host of new small bars in all shapes and sizes, in weights from 1 gram up. The vogue was pioneered by Credit Suisse, whose Valcambi refinery at Chiasso in Switzerland turn out an exceptional high-quality range of baby bars. They proved immensely popular in the Middle East (which has plenty of locally made coin copies but has never been penetrated by bullion coins) and in Singapore. For a while, the Credit Suisse branch on Shenton Way in Singapore became a Mecca for Singapore's gold investors, who queued up to buy and sell at special gold windows in the bank. Prime Minister Lee Kuan Yew once confided to a friend that he used to judge the state of play in gold just by the length of the line outside Credit Suisse when he drove by.

The small ingot fashion was later expanded by Mahmoud Shakarchi, Beirut's best-known gold and exchange dealer for many years until the civil war drove him and his family to Switzerland. Shakarchi established the Pamp refinery, also in Chiasso, initially in a single room but later in a large new factory, and began producing a galaxy of small bars. The most famous is the 'Fortuna', depicting the Roman goddess of Fortune emerging from a conch shell (symbol of plenty). The Fortuna, in sizes from 1 gram up to 50 grams, was an instant hit throughout the Middle East, achieving sales in the best years of nearly 30 tonnes (almost 1 million ounces). Its secret was that it was both an investment and a jewellery item, selling on an exceptionally low mark-up. Variations on the Fortuna theme have been endless, with Pamp turning out pear-shaped bars one year, diamond-shaped another. And their success has promoted other refineries to follow them.

Although the novelty value has worn off and the Middle East itself is buying much less gold, the small bar market is an important niche, taking up anything between 25-50 tonnes annually. The difficulty facing the small bar, compared to the bullion coin, is that in some markets, such as the United States, it is less of a two-way vehicle. Selling back in Middle East *souks* is no problem, but in New York or Los Angeles it is much more difficult.

Yet, given the challenge facing the gold mining industry to lift the 'safety net' for the price in years when many investors neglect gold, the small bar sector, along with coin, has potential for growth. Silver mining companies have often offered their shareholders silver bars. Why should not gold mining companies offer dividends in their own small bars? The point is that bullion coins and small bars have established a firm base in the gold business. Taken together they can absorb, at worst, 150 tonnes a year and, at best, 350 tonnes. They represent the best tranche of the market, except jewellery, to build on.

Portfolio Fireworks

Fireworks in gold rarely come from the daily physical market. A real display can be provided only by speculative foray, which must be coupled with a sustained shift of investment funds if it is to be maintained. Even the physical investment we have been discussing in this chapter provides only the long-term cushion for the price. Gold has to tap the extravagant amounts of money with which the currency, stock and bonds markets have been awash for several years if it is to climb to $1,000 or $2,000 an ounce. When the dollar first came off its peak in 1985, a banker in Geneva said to me, 'There's a huge sea of dollars out there, but it is starting to drain out into rivers and streams, and one of those rivers will be precious metals. We are starting to make sure we have four to five per cent in gold in the portfolios.' As it happened, he was a trifle early in his prediction, because the stock markets took the speculative flow for a while, but his basic instinct was correct. Only a small diversion into precious metals was necessary to consolidate the price.

Twice in the last twenty years, private investors have shown their ability to soak up tonnes of gold. First, in 1968, when the central banks' gold pool collapsed against a determined private onslaught brought on by the Tet offensive in Vietnam. In a single week, the pool put out 1,000 tonnes from official reserves to try to hold the price at the old $35, and then gave up. Then again, over 1,300 tonnes of gold sales by the IMF and the US Treasury did nothing to halt the price advance between 1976 and 1980. The lesson from these episodes is simple. If the investor has a mind to do so, he can take up all the gold thrown at him.

This is particularly true if two of the three strands of investment demand identified above pull together simultaneously - namely, portfolio investment coupled with a strong bullion coin demand for the small investor, with perhaps the added bonus of good physical kilo bar demand in places like Japan (or the Middle East in earlier years). This happened from 1978 to 1980, when the combined forces of portfolio, coin and bar hoarding provided an exceptionally strong undertow, taking up nearly forty per cent of all gold coming to the market. In 1980 alone, with the price averaging over $600, the combined investment sector took up fifty-six per cent.

Since then the marketplace has evolved substantially (see Chapter 11), so that in future it will be tapping a broader segment of investment money. The remark of Jeffrey Nichols of American Precious Metal Advisers, that the growth of infrastructure of the US market has made it easier for a larger audience of mainstream investors to participate in gold, is worth remembering. In the last bull market to 1980, the American investors were much less visible than those from the Middle East, who were the real driving force. But the Middle East, as we have seen, is a shadow of its former self, and although a good deal of money can be directed into gold, particularly through managed funds from the region, the speculative sparkle has gone. The future lies much more with the Americans and Japanese. If they really get their teeth into gold, then the price could go to levels beyond those anticipated by even some of the most optimistic forecasters.

Looking into the crystal ball for the gold price, however, it is important to detect the right signals. A short-term speculative run, as happened in late 1982 or the autumn of 1986, should not be mistaken for a genuine shift of investment sentiment which can give a sustained lift to the price. On both those occasions, the price surge was essentially technical through the covering of short positions of speculators caught out on a limb. The real sea change comes only when portfolio managers decide to go into gold (or, later, to get out).

The best signals are in Switzerland, especially among the Geneva private banks, whose antennae are subtly tuned to pick up the first murmurs that it is time for a change. Their investment strategy committees do not take decisions lightly, and do not chop and change. If they decide to increase portfolio exposure in gold to, say, five per cent, that will be the policy for some months, if not for two or three years. They will also go up by gentle stages, perhaps five per cent, then seven per cent, then ten per cent, and even to fifteen per cent in the right context (such as 1979-80).

These decisions are the ones that have lasting impact upon the development of the gold price, because it takes a while to build up that exposure, buying carefully in any dips. This provides the market with good support, but is entirely different from chasing the price up in a speculative foray.

The amount of money under management in Switzerland alone is such that a shift of opinion can involve an engagement in gold that makes any normal supply-demand equations irrelevant. The twenty-four private Swiss banks were alone estimated to be managing funds valued at SwFr150 billion (around $100 billion) at the end of 1986. Thus, if they all decided on ten per cent in gold, this would imply investment of $10 billion in the metal, requiring 777 tonnes with gold at $400. That is for the private banks alone; almost the tip of the iceberg. Another estimate indicates that Zurich banks, including the big three, have $500 billion under management; ten per cent of that would take up two years' new supply of gold.

The perception of gold in Switzerland, however, has changed, as elsewhere. The classic Swiss portfolio in the days of a fixed gold price was ten per cent; the gold was regarded not as a performer in the portfolio, but as insurance. With a floating gold price, the attitude is different; gold must earn its keep and be managed profitably like any other asset. Thus, throughout the 1980s, exposure to gold was often run down to very low levels - almost nil in some circumstances. Only in the summer of 1986 did strategy committees again start to make sure they had four or five per cent in gold. By mid-1987 that had often been raised to ten per cent. I asked one private banker when his clients had last been recommended to hold that much gold. 'In 1979-1980,' he said. It was that policy change which gave the gold price solidity over $400 in 1987. Japan shouldered the burden in 1986, the portfolio investor picked it up thereafter.

This portfolio involvement with gold, however, still tends to be on behalf of very wealthy private clients and only to a small extent on the part of the big institutional investor, especially in America. Many of them still remain at arm's length from gold, either prohibited by law from engaging in precious metals or by the natural inhibition of money managers, who are judging so closely on performance every two or three months that they would prefer to stay with bonds or equities which they feel they can juggle to show a good profit. Pension funds, for instance, remain wary on gold, even in Switzerland. 'There is tremendous potential in pension funds,' a Zurich banker told me, 'but they are very conservative. I've managed to persuade one to give me 10 million Swiss francs (about $6.5 million) to try in gold. But it is slow.'

At best, the institutional investors and pension funds play it safer and go for gold shares, either directly with individual mines or mining houses, or through the increasing variety of gold funds. These funds tend to be heavily into gold shares rather than bullion itself. (In Britain, gold funds are not allowed to go into the metal.) North American gold funds, in particular, have been attracting much more money. At the end of March 1987, twenty-four listed US gold funds had assets totalling $4.2 billion, compared to $1.6 billion just fifteen months earlier. But the major slice was in gold shares. 'We're not yet so metal-orientated in the US,' an analyst at Goldman Sachs, the New York banking house, told me. One hurdle is that some funds are not permitted to have more than ten per cent of their investment in 'unusual' assets, which still include gold. 'Most funds are trying to figure out a way to get around that and buy more metal,' the analyst went on, 'because it is more liquid than the shares in some of the smaller mining companies.' But what they will often do is 'park' new inflows of money in gold itself, while waiting to get the right shares.

Things are easier in Canada, where the four major gold funds, BGR Precious Metals Inc., Central Fund, Goldcorp and Guardian-Morton Shulman, with combined assets of C$480 million have all been able to go for higher bullion (gold and silver) exposure. Central Fund, for instance, has consistently maintained well over ninety per cent in gold and silver, earning income, too, by writing options on the holdings.

The underlying potential for all North American funds is substantial as they start to attract the attention of a new generation of money managers with little knowledge of gold or gold shares. A mining company executive in New York recalled, 'At a lunch the other day I sat next to the manager of the investment board handling the pension fund for a Mid-Western state. He had come to learn about gold, to which his funds had virtually no exposure. He needed to spend $150 million just to bring his fund up to the 8.5 per cent exposure he had in mind.' Although that state pension money will first go into gold shares, the next step could be the metal itself. Gold funds with a real bullion content have great appeal. When Swiss Volksbank launched a new gold fund in April 1987, it took in SwFr200 ($133 million) in just three days.

Gold Bonds – the Dark Horse

The dark horse creeping up in the investment game is the gold-linked bond, which suddenly jumped into fashion in 1987 on renewed fears of inflation. The initial play with gold bonds had already come from gold mining companies, linking gold warrant calls on their production as a way of raising money. Banque Paribas Capital Markets, for example, launched a bond for Canada's International Corona Resources that is convertible into gold. And Bank Gutzwiller, Kurz, Bungener (Overseas) Ltd, led a similar 7.5 per cent bond for the American producer Pegasus Gold Inc. The success of these bonds prompted others to use the same concept to raise money for quite different purposes as expectations of better precious metal prices grew. 'The advantage is that you pay less interest, but offer the attraction of the gold price,' said a Zurich banker.

Three Swiss franc bonds were launched, for example, by Credit Suisse for Electricite de France, for the Belgian government and for Hoffman-La Roche, the Swiss chemicals and pharmaceuticals group, with warrants providing a claim on Credit Suisse's own gold. Salamon Brothers then pitched in with a 75 million-ecu issue for St Gobain Netherlands, paying 4.5 per cent interest and carrying warrants allowing holders of each 1,000-ecu bond to buy an ounce of gold at $490. Citibank NA Zurich came up with 20,000 warrants at SwFr610 each, entitling the holder to buy gold at $440 an ounce. Although these gold commitments can be hedged with options, they provide a new underlying bullish factor in the gold market. 'There must be some gold put aside to cover the warrants,' said the Zurich banker, 'Not a huge amount perhaps, but count a few tonnes.' Another banker in Hong Kong was more enthusiastic about the prospects. 'I'm convinced big dollars could flow into gold bond issues in the capital markets,' he said. 'If we could issue $100 million gold-indexed bond tailor-made for institutions, we would start to turn a commodity into a security.'

The gold bond was particularly topical in 1987 because of the maturity on 1 January 1988 of the French government's bond 'Rente Giscard'. This bond, issued in 1973, carried a guarantee that both the seven per cent interest and the maturity value of the bond would be indexed to the price of a 1-kilo bar on the Paris Bourse if the official link between the value of the French franc and gold had been severed in the meantime. Precisely that happened in 1978, when the IMF agreed that the currencies of its members were no longer gold-backed. So the maturity value of the bond itself will be determined by the average price of gold in Paris over thirty trading sessions prior to 1 January 1988. The French government will have to set aside over $8 billion to redeem the bond. If bond holders demand all that in metal, it could involve 600-650 tonnes, if the gold price was between $400 and $450. The French have the gold, of course, but it is nearly twenty-five per cent of their official reserves, which they would be reluctant to diminish. The French government will naturally seek to roll the debt over, with a new gold-linked package. But some bond-holders may demand gold, and it is not impossible that 100-200 tonnes will be called upon. That could well be covered in the market. An added dimension is that many bond-holders bought the Giscard bonds when they were standing at a discount of up to fifty per cent. They financed these purchases by selling borrowed gold but before January 1988 will have to acquire gold to repay their borrowings. They may have to buy quite large amounts in the open market to cover. So the denouncement of the 'Rente Giscard' has given an intriguing twist to the blossoming gold bond scene.

The difficulty in all this is to keep a sense of proportion. A scenario can be conjured up showing gold bonds, gold funds and general gold investment suddenly transforming the outlook for a market that previously seemed all too bearish. The volatility of sentiment was summed up by a London analyst who said, 'In a bear market, everyone says, How can all this new production be absorbed? In a bull market they say, How tiny the gold market is compared to all that money seeking a home.' That is the dilemma for the gold miner. The investor can carry the day when the mood takes him. But what about the years when it does not? The challenge for the gold mining industry is to raise the safety net of physical offtake for moments when the investor goes elsewhere. The keynote then is jewellery.

Chapter 14

THE FABRICATORS: JEWELLERY, THE ERODING CORNERSTONE?

The crossroads of the jewellery business each spring is Basel in Switzerland. Manufacturers, designers, wholesalers and retailers from throughout the world gather together in the Messeplatz Exhibition Centre for the European Watch, Clock and Jewellery Fair. There are, of course, other jewellery fairs, at Vicenza in Italy in January and June, at Munich in February, and at Valencia, Spain, in April. But Basel is unique. Skim through the catalogue: 436 manufacturers of gold jewellery, fifty-seven manufacturers of gold watch cases and twenty-seven manufacturers of gold watch bracelets turned up in 1987 to display their wares. They are a Who's Who of gold: 'Balestra, 1882, catena, collieri, bracciali, annelli in oro 18 ct'; 'Caplain Saint Andre, depuis 1823, le marque d'or, Paris'; 'Chopard, Geneva bijouterie ou orfeverie en or'; Louis Fiessler, Pforzheim; Chap Mai Jewellery, Singapore; Chin Bros Arts Co., Hong Kong; Gold of Jerusalem, Israel; Kaneko, 'jewellery woven with romance', Japan; Lapporiia Jewelry OY, Finland.

The Basel Fair is the place to take the pulse of the gold jewellery industry. 'The US is overstocked, they are up to here,' comments an Italian chain-maker, covering his eyes. 'The Middle East is more positive, picking up 20-25 gram items,' says another. 'The Italians are taking off for Japan, because the US is slow,' remarks a wholesaler. A manufacturer from Pforzheim, the home of West German jewellery, laments that high street chain stores are replacing the traditional family jeweller. 'There is no engagement of an owner selling from his heart,' he complains.

Jewellery is the cornerstone of the gold business. In the years 1968-86, 14,462 tonnes of gold was fabricated into carat jewellery, out of 25,396 tonnes of new gold supplies available to the private sector (that is after netting off scrap and central bank transactions). Jewellery in this period has consumed virtually fifty-seven per cent of all gold supplies. This includes jewellery for the Middle East and South-East Asia, which is bought both for investment and adornment, as we noted in the previous chapter, but is an integral part of overall jewellery demand which, year in, year out, underpins the market. The long-term outlook for gold has to consider, above all, whether the jewellery sector is keeping pace with the rapid increase in production. The short answer is 'no'. Jewellery's contribution is being slowly eroded. Throughout the 1970s, sixty-five per cent of the gold coming to the market (including US Treasury and IMF sales) went into jewellery; in the 1980s it is down on average to fifty-one per cent (excluding the use of recycled gold).

The jewellery trade was exceptionally hard hit by the high price of 1980. Fabrication from new supplies tumbled in that year to a mere 280 tonnes, compared to over 1,000 tonnes annually in the late 1970s. It has never fully recovered. The best year since was 1985 with jewellery at just under 900 tonnes. The following year, with the gold price higher and the Middle East in a liquidity shortage, it was down to 827 tonnes, or forty-nine per cent of new supplies.

A distinction between new supplies and recycled scrap is vital in any long-term analysis because in many countries, including, for instance, Egypt, India, Indonesia, Saudi Arabia and several Latin American countries, a good deal of fabrication is from local recycling at all times, and in periods of high prices may be entirely from that source. An estimate, therefore, of total jewellery use from new and old gold can give a deceptive figure of how well jewellery is really doing. Local retail sales may be steady, but of jewellery made entirely from scrap. The picture in 1986 is an illustration: fabrication including scrap fell by only 2.5 per cent, but fabrication from 'new' gold declined 7.2 per cent. Since gold supplies from mines and Communist sales were up, the amount of metal absorbed by jewellery was clearly not keeping pace.

Is the cornerstone showing signs of collapse? We have already observed, in the previous chapter, how the investment jewellery demand in the Middle East and South-East Asia is very price-sensitive and, coupled with reduced purchasing power in OPEC countries, has been declining. The drop is not just in local fabrication in countries as varied as Egypt, the Gulf States, Singapore or Indonesia, but in exports from Italy (and, to a smaller extent, from other European centres as well). Italian exports to Saudi Arabia fell by fifty per cent, and to the United Arab Emirates and Kuwait by over sixty per cent between 1984 and 1986.

High Mark-up, Low Sales

The situation in Europe, North America and Latin America does not show such dramatic decline (Panama, for instance, is now Italy's third best export customers, being the wholesaler for Latin America), but it is scarcely improving. The whole concept of jewellery, however, is different; it is bought principally as a fashion item for adornment and not for investment. The investment motive may linger in some people's minds, but the price paid rarely justifies it. The contrast with the Middle East is stark. There, investment jewellery is usually 21 or 22 carat, and is sold on a low mark-up of ten to twenty per cent over the gold price of the day to cover the 'making' charges and a small margin for the retailer. But in Europe, the United States, Latin America and Japan, mark-ups may be anything from 150 to 400 per cent, while the gold content is a mere 8 or 9 carat. (8 carat means the jewellery is only one-third gold alloyed with other metals, such as silver or copper.) In Britain, for example, over ninety per cent of jewellery is only 9 carat (despite frequently being advertised as 'solid' gold). Only in France, Italy, Spain and Switzerland is 18 carat the minimum that may be described as gold jewellery. And only in Italy and Spain, where an investment motive lingers more strongly, will mark-ups be low. As a result of that, Italians buy about two and a half times as much jewellery per head as the British or the French.

Undoubtedly the high mark-ups on low-carat jewellery kill the market. The normal buyer goes into a jewellery shop with between $200 and $500 to spend. If the mark-up is 400 per cent, he gets only $25 in pure gold for every $100. That means, almost inevitably, that the item itself will have to be light to fit his budget, usually 5 grams or less. No wonder Italian manufacturers talk about 'fly-away' chain for the European market that is so light it will blow away if the window is opened.

The high mark-ups (Americans call doubling the price from factory to wholesaler and wholesaler to retailer 'keystoning') are really an inheritance from $35-an-ounce gold. Labour charges were then seventy per cent of the cost of an article, so mark-ups were justified. But, with gold at $400 or $500, the relationship is reversed: the gold itself is the prime cost. Yet the jewellery trade goes on keystoning. The concept of 'making' charges has not been widely adopted.

The Italian Connection

The result has been declining or, at best, stagnant, jewellery sales in most developed countries for the last decade, in terms of tonnes of gold consumed, although not in money spent. In Western Europe, for instance, jewellery fabrication in 1986 was thirty per cent less than ten years earlier, if Italy, the workshop for the world, is excluded. The only sustained improvement has been in Switzerland, through the buoyancy of the gold watch case and bracelet business, in which fabrication has virtually doubled. Otherwise only in Italy has the fabrication level held up well, through the wizardry of the great chain-making companies like Gori & Zucchi (Uno a Erre), Balestra and Oromechannica, offering highly competitive making charges and securing export contracts to every continent. 'We work with twenty-one markets constantly,' said Patrizio Taddia, sales director of Balestra, 'Europe from Sweden to Spain, Middle East, America, South America, Korea, even a little to China now.' Despite exports, Italian fabrication in the 1980s has never quite equalled the peak year of 1978 when 235 tonnes of gold were transformed into jewellery; the best since then was 231 tonnes in 1985.

Italy's fortunes for almost a decade were founded on the Middle East, but by 1986, less than twenty per cent of her exports went there. The United States became the best customer during the years of the strong dollar, taking over forty per cent of Italian production from 1984-86. But the dollar's demise has eaten into that. At the Basel Fair in 1987 many Americans were conspicuously absent.

'This year is lost,' said an Italian chain-maker. 'We're tailored for the US market, and now it doesn't exist. We do not have enough flexibility, we have not diversified the risk. We'll have to change our machines and approach the Japanese market, but that takes time.' The Japanese themselves are showing enthusiasm. 'We have 500 Japanese at the Vicenza Fair, and at Basel it's rumoured there are 4,000 of them,' said another Italian fabricator. But what worries the Italians is whether the Japanese will really become large customers or just learn the techniques themselves. 'The Japanese come to our factories and take notes about everything,' he went on. 'The Americans tried to copy us, but couldn't do it, but the Japanese might.'

Japan's potential is magnified by Europe's lack of enthusiasm. 'In Europe we have to compete with new colour TVs, videos and holidays abroad,' said Balestra's Taddia. 'People say I've bought two pieces of jewellery as fashion, why should I keep on buying gold? I'll go to the Seychelles instead.' The high mark-ups in Europe also make it difficult to go on selling expensive gold jewellery into a fashion-conscious market that loves new designs every season; costume jewellery fills that bill more cheaply. 'A piece of gold jewellery may cost 8,000 deutschmarks, but a similar costume ornament is only Dm800,' said Dr Gert Furhmann of Louis Fiessler, the Pforzheim manufacturers of high-quality carat gold. 'People find the price too high, so they buy costume jewellery.' The answer, some Italians argue, is that their industry must think less of volume and more of better-quality, personalised products designed for specific markets. They have been able to coast for nearly twenty years first on a unique market based on the Middle East's oil boom and then on the United States. Wholesalers in both were more interested in the price per kilo than the price per item. 'Italy is changing already in terms of quality,' said Taddia. 'The machine-made chain by volume will always be needed, but new merchandise will give us more profit margins to diversify.' I asked Taddia what that implied for gold consumption in Italy over the next decade. He was quite specific: 'More quality, less quantity.' That may augur well for the profitability of the Italian industry, but not for actual gold consumption. Italy has accounted for one quarter of all jewellery fabricated from new supplies during the 1980s; in its turn, it is the cornerstone of gold jewellery. A decline in Italy takes a lot of replacing elsewhere. Not that Italy's eclipse is imminent. The tradition and expertise in the towns of Arezzo, Vicenza, Bassano del Grappa and Valenza, Po, is not seriously challenged, but their adjustment to new trends is likely to hold fabrication at around 200 tonnes annually, rather than exceeding the 235 tonnes of the late 1970s.

While changing styles, the Italians are seeking fresh markets. Eastern Europe looks promising. 'The people have money, but nothing to buy in the shops, so there is jewellery smuggling,' confided a chain-maker. 'I know, because I do it. We are getting orders for many thousands of 14-carat items, sometimes a hundred kilos at a time.'

Salvation in Japan?

The real promise, however, is in the Far East. Jewellery consumption in Japan rose by nearly thirty per cent in 1986 alone. Most of this was accounted for by local fabrication, but imports from Italy, Hong Kong and Singapore also edged up. Japan is starting to assume the importance in gold jewellery that it has already in coin and investment. Gold chain is the most popular item, especially a distinctive local variety of 18-carat curb chain known as 'kihei', that is sold by weight. Kihei chain accounted for over thirty per cent of all 87 tonnes of jewellery sold in Japan in 1986. Outside Italy, Japan now has the best chain-making facilities, with over forty companies producing machine-made chain. The two largest, which account for fifty per cent of fabrication, already rival Italy's largest in gold use. The strong yen makes exports difficult, but it will be only a matter of time before Japan's chain-makers follow their colleagues in cars and electronics into world markets.

The Japanese, incidentally, also find extravagant uses for gold on a scale only equalled in the past by the golden dinner services made for Indian maharajahs. Companies present favoured customers with 24-carat teapots, vases, chopsticks or saki cups to honour important contracts. At a hot spring outside Tokyo, visitors can take a plunge in a phoenix- shaped tub fashioned from 143 kilos (4,580 ounces) of pure gold. The latest attraction opened in May 1987 is a golden tea ceremony room in the Moa Art Museum at Atami in Shizuoka prefecture, using 50 kilos of gold. The pillars and beams of the room are covered with rolled gold and the walls are papered in gold leaf. The teapots and cups have been made from 24-carat gold. The Precious Metals Market survey produced by Sumitomo Corporation accounts for over 3 tonnes of gold required annually for such 'artistic ornaments'.

Chuk Kam Challenge

On a simpler scale, China is also seen by many fabricators as the best untapped market in Asia. The open door policy in China has already enabled jewellery fabricators to get an informal foot in the door. The Chinese government has permitted jewellery sales to be resumed since 1982, and they already account, according to official figures, for $500 million annually. Most of that jewellery, however, is home-made. What may become more significant is the flow of jewellery into China with travellers going to visit family and friends. A sharp rise in jewellery fabrication in Hong Kong in recent years is already attributable in part to this traffic. At Chinese New Year, for instance, over half a million Chinese from the colony visit China. They take with them consumer goods, including the popular 'chuk kam', or pure gold, jewellery that has become the vogue in Hong Kong. Chuk kam bracelets and necklaces, often weighing 25-50 grams, are classic items of investment jewellery, sold on low mark-ups through the shops of the big Hong Kong jewellery groups, Chow Sang Sang, Chow Tai Fook and King Fook (which also has a shop on Fifth Avenue in New York).

The success of chuk kam in Hong Kong has stimulated a technical breakthrough that could have immense significance for the future of the jewellery industry if it is widely adopted. Chuk kam must be a minimum of 990 fine according to local hall-marking rules to be billed as pure gold, a quality that is really too soft for durability as jewellery. Experiments by the Research Institute for Precious Metals and Metals Chemistry at Schwabish Gmund in West Germany have shown that, by alloying a touch of titanium with gold, remarkable durability can be achieved. It is as if the titanium provides the gold with a tough skin. This discovery, if it can be adopted commercially, has tremendous implications for the jewellery industry. The concept of chuk kam, still limited mainly to Hong Kong but now also being test-marketed in Japan, is an essential step along the road to wider acceptance of high-carat, low mark-up jewellery. The most positive step that could be taken to increase gold consumption in jewellery would be to wean the public away from 8, 9 or even 14-carat gold to higher caratages. The dilution of gold with other alloys has progressed too far. Jewellery is too lightweight, too anaemic. A shift back to the rich golds of heavy 21 or 22-carat necklaces and bracelets would be a tonic for gold consumption, and quite possibly a pleasant surprise for buyers.

The effect upon Europeans and Americans coming across high-carat, low mark-up jewellery for the first time can be astonishing. In the souks of Saudi Arabia and the Gulf in the evening the gold shops are crowded with American, English, French or German expatriates buying with an eagerness they would never show at home. For the first time, they find they can buy a chain necklace of 30-40 grams or a bracelet of 50 grams almost at the gold price of the day. In twenty years of visiting the Middle East I have never known a colleague, man or woman, going there for the first time who did not buy at least 1 ounce of gold jewellery. And it then becomes a regular habit. But do they ever buy an ounce of gold jewellery at home? Never. It takes a special event to lure them over the threshold of a jeweller's shop. In the souks they find a social centre where gold looks like gold and the price is cheap.

The response of most Western jewellers to suggestions that they, too, opt for high-carat, low mark-up is that their jewellery is more finely crafted, that they have higher overheads, and have to pay taxes. That is true, to a degree. But a compromise between ten per cent and 400 per cent mark-up ought to be possible. The object lesson, in fact, is in the Asian communities of Britain, where immigrant craftsmen have carried on making 22-carat rings, bangles, necklaces, pendants and earrings to be sold for ten per cent mark-up, plus labour charges of about ten per cent, plus value added tax. Hallmarking figures reveal 22-carat jewellery as the fastest growth sector in Britain; hallmarking of such items jumped from 1.8 tonnes in 1980 to over 4 tonnes by 1985, and now represents twenty per cent of all jewellery hallmarked in Britain. Naturally, this disproportionate level of gold jewellery acquired by a tiny segment of the population has much to do with their social customs, but the fact that one of the liveliest gold jewellery centres in the world has developed in north-west London shows that the argument about overheads holds little water if the right mood is there.

The Innovators

The message has not gone unnoticed. A few entrepreneurs are pioneering avenues that are essential if jewellery fabrication is to hold its place as the cornerstone of the gold business.

The inspiration, not surprisingly, has been the Far East. A young Texan named Seth Hersh who worked for several years in Indonesia was captivated by the heavy 22-carat bracelets and necklaces handmade by local craftsmen. He went back to Dallas and established a business called Batikat to import and market them, on low mark-ups of fifty to sixty per cent. I first encountered Hersh at an investment conference, where he had a small display counter in the exhibition centre. All round were other jewellery retailers offering 14-carat gold chain by weight, which sounded tempting until the mark-up was calculated - it was the normal 400 per cent. Seth Hersh simply offered 22-carat necklaces weighing 30, 50 or 100 grams. When he first placed them in people's hands they were amazed at the heaviness; when they found the price was around $600 an ounce, with the gold price at $400, they bought. Hersh's philosophy is simple. 'By lowering the traditional mark-ups,' he said, 'the goldwork is truly of considerable value to the customer.' He encourages them to pay in gold, and offers, too, a guaranteed buyback in gold. 'Each 100-gram necklace - that's just over 3 ounces - will retail for 5 ounces of gold coin,' he explained, 'and it can be sold back to Batikat at a price equivalent to 3 ounces.' He also quotes prices to wholesalers and bullion dealers (who market much of his jewellery) in gold coin. On the 100-gram necklaces, he charges a wholesaler ordering at least twenty pieces only 3.75 ounces of gold coin, a mark-up over gold content of seventeen per cent; a far cry from conventional 'keystoning'.

Batikat is a trail-blazer for the high-carat, low mark-up concept in North America. In itself it cannot transform the market, but the interesting thing is that once people have discovered the idea they forsake the customary 14 carat. And they start to buy 1 or 2 or 3 ounces of gold jewellery at a time, just as happens in the Middle East, South-East Asia or even in Britain's Asian communities.

The other hurdle to be overcome is to lure customers into jewellery stores, not just when they are getting married or are duty bound to buy a gift, but to browse round as they might in a bookstore and perhaps buy on impulse. Most people simply do not wander casually into a jewellery shop, because they fear being confronted instantly by a salesman out to sell them an expensive bauble. That is not the problem in the Middle East *souks* where the gold shops are social centres and customers sip tea, coffee or cold drinks. Nor is it at such showplaces as the Yamazaki store on Ginza in Tokyo, which is the ultimate gold shop with customers wooed by pop music and video fashion shows as they step off the street. The idea is taking off elsewhere.

In Paris, stroll down the fashionable Rue Auber from Place de L'Opera and the entrance to a very bright, white jewellery store Ora Bora beckons. There are no doors to open, and a cascade of jewellery is displayed on neat racks in alcoves all round the walls with prices clearly marked. It is more like going round an art exhibition. Ora Bora (which has two shops in Paris) is the creation of Bernard Denet, who worked in the jewellery business in Hong Kong and Japan for several years and, like Seth Hersh, became fascinated by the possibilities of introducing an Eastern approach to jewellery sales. He has put less emphasis on high carat (the French, anyway, accept nothing less than 18 carat) and low mark-ups (not easy with the overheads of a fashionable Paris street) than on style and presentation. 'We get over the threshold problem,' he told me. 'You walk into a relaxed jewellery environment. It's all clearly displayed and we give you five minutes freedom to walk around before a sales lady quietly approaches.' The significance of Ora Bora was not so much its own sales, but that it made the jewellery trade in Europe take notice. 'We proved it is possible to do something different,' said Denet. 'We have three jewellery chains in France copying us, and one German chain sent three architects to look at the shop and then opened one which is a straight copy.'

The Gold Promotors

The need for new directions and the need to underwrite future jewellery demand is finally getting through to the gold mining industry.

Although the International Gold Corporation, supported by South Africa's

Chamber of Mines, has been promoting gold for over a decade, the industry as a whole was not interested. This neglect was pointed out at an international conference by Julian Baring of James Capel, the London stockbrokers, who noted that the soap powder industry spends six per cent of its turn-over on advertising, but the gold mining industry scarcely 0.4 per cent. 'It's my belief that the gold mining industry is in danger of allowing complacency to overtake it,' said Baring, 'at a time when the profit margin remains high enough to attract so many newcomers to the industry. When you think what De Beers has achieved with diamonds, it makes you wonder why the same effort is not made with gold.'[1]

His gauntlet had, in fact, already been picked up. A month later executives from Australian, Brazilian, Canadian, South African and US mines met in San Francisco to confirm the launching of the World Gold Council (WGC) to present a united front in gold promotion. The initial budget for the WGC is just over $60 million annually (about half of what De Beers spends on diamonds) and the aim is $90 million. That depends, however, on how many mining companies can be persuaded to join the sixty-two founder members from seven countries. The WGC is planning to spend seventy per cent of its budget promoting jewellery, with the balance encouraging investment and industrial applications. The proportion devoted to jewellery, which is likely to be channelled principally into television commercials, acknowledges that for the long haul this sector must be the target for growth. Jewellery has to be restored and maintained as the cornerstone.

Television commercials, however, are not an elixir. Just as much energy, if not money, needs to go into coaxing the jewellery trade along the avenues of higher caratage, lower mark-ups and a friendlier welcome to customers. Bernard Denet's plea for a 'relaxed environment' is important. So are initiatives on new products. The small-bar boom, which we discussed in the previous chapter, put 5 and 10-gram gold bars suspended on chain round the necks of secretaries and Wimbledon tennis players. On Mediterranean beaches it is all some people wear. The baby ingot really must be counted as jewellery.

Some initiatives come almost by chance. A young American, David Santomenno, went on vacation to Panama a few years ago, and on the first day in the Museum of Panamanian Man saw golden 'huacas' of beetles, butterflies, crocodiles, frogs and turtles fashioned by pre-Colombian Indians six or seven hundred years ago. In an instant, he decided to start a workshop re-creating them. 'I had one day's vacation and then started working,' he recalled, 'I had no idea of pre-Colombian artifacts or of the lost wax technique by which they were made. But I persevered and mastered it. It's amazing that Indians with no technology did all this.' At Santomenno's little factory, christened Reprosa, in Panama City I watched craftsmen making 14 and 18-carat replicas of 'the virility frog', regarded by the Indians as the bringer of new life as it searched for newly

1 Julian Baring, *Financial Times*, World Gold in 1986 Conference, London, 18 June 1986

formed pools in damp undergrowth, and the jaguar, who was their strong and powerful god. An assortment of Reprosa's *huacas* adorns the cover of Panama's telephone directory, and they are sold at gift shops in American museums and through the American Archaeological Association. The original splendour of pre-Colombian gold is best seen in the Museo del Oro at Banco de la Republica in Bogota. The bank started collecting in 1939 and now has a unique collection of 10,000 objects. They, too, have inspired a Colombian goldsmith, Guillermo Cano, to develop a successful export business in pre-Colombian reproductions. Archaeological inspiration is not limited to Latin America. In Athens, Ilias Lalaounnis has won a considerable reputation in re-creating ancient Greek jewellery found in the excavations of Minoan Crete and Mycenae.

The Golden Orchid

The amount of gold used by such craftsmen is small, but it offers diversity from conventional jewellery and taps a new market. Reprosa in Panama, building on its reputation with *huacas*, is using the same lost wax techniques to create golden replicas of orchids and African violets, cast from original flowers. They sell the golden flowers to collectors at international orchid conferences and through publications like the *African Violet Review* (which has a substantial following). In rather different style, Bernard Passmore, a sculptor in the Cayman Islands, uses gold offset against highly polished black coral to make all kinds of trinkets from miniature golf clubs to scuba divers, which he sells to the wealthy at Florida resorts.

Broadening the demand for fabricated gold, however, must be coupled with a long-term determination to up caratage and cut margins. The mining industry has agonised for years on whether vertical integration is the answer. Should the mines own jewellery factories or retail outlets, cutting out the traditional intermediaries? They have hesitated, primarily for the good reason that they know how to get gold out of the ground, not how to design and market jewellery. But the pressures to sell more gold could still lead them in that direction. Some bullion dealers argue it is the best solution. A leading European dealer put it bluntly: 'I'm doubtful if advertising will help. The mines need more entrepreneurial skill. They should have half a ton of gold made into jewellery, distribute it to large department stores on consignment and tell them to cut margins. Run a test and see how it works. They'll lose lots of friends in the jewellery trade, who'll cry like hell, but in the long run if the mines don't achieve use of metal in jewellery and industry, I don't see any chance of a sustained price rise for an extended period. Investors can help from time to time, driving up the price, but then we're in for long period of deteriorating prices.'

In short, bold decisions are required to reconstruct the cornerstone.

Industry: Nothing is as Good as Gold

The phrase 'nothing is as good as gold' is often used to tout its investment virtues. It really describes more aptly its industrial applications. Nothing is as good when it comes to malleability, ductility, conduct of electricity, resistance to corrosion and reflection of heat. Gold bonding wire filters electricity through integrated circuits in home computers and television sets; gold plating of connectors and switches on undersea cables or satellites ensures long-lasting reliability; gold-coated glass keeps buildings cool and protects astronauts from the sun's harmful heat rays in space; and dental gold alloys offer excellent protection from corrosion in teeth.

Such testimonials ought to ensure that here at least gold use is improving. On the contrary, overall it is stagnant and in such applications as dentistry is declining quite sharply. Gold consumption in industrial, dental and decorative uses has hovered between 200 and 230 tonnes annually for almost twenty years. Dental use has actually fallen from a peak of 93 tonnes in 1978 to around 50 tonnes, but that has been offset by a strong electronics performance in the 1980s. Salvation has come from the integrated circuit, a micro-electronic web of hair-like gold wires spanning the surface of a silicon chip, that is at the heart of new technology in everything from video recorders to cars and military hardware. Gold's ductility has enabled the wire for the integrated circuits to be drawn to a thinness of 25 thousandths of a millimetre. Japan has carved out the lead in this new era of electronics. She has forged ahead of the United States as the largest fabricator, accounting for forty-two per cent of the 123 tonnes of gold used in electronics in 1986.

Japan's success, however, masks intense research effort not just to use less gold in each circuit, switch or connector, but to replace it with other metals, like palladium nickel and silver. Already the thickness of gold-plating solutions on connectors has been reduced to the thinnest of films, as if replacing a thick winter coat with a thin summer dress. The gold plating is also 'spotted' at strategic points of the connector, instead of being spread all over it. So far, this has not reduced overall consumption because the electronics industry has grown so fast it has required more units. The drive to find alternatives to gold, however, is becoming a serious challenge. 'In the realm of electrical connectors, gold and its alloys are under continuing threat from palladium-nickel layers finished with a flash of gold, palladium-silver and nickel with a small addition of phosphorus,' reports Dr George Gafner, research and development director of the World Gold Council. 'Some of these substitutes are remarkably good and will undoubtedly make long-term inroads into the use of gold in industry.'[2]

His opinion was echoed in Japan by Tanaka Kikinzoku Kogyo, the leading

2. Dr George Gafner, *The Significance of Gold in Industry*, Gold Update, April 1987

precious metal fabricators, when I asked them about the industrial view to the year 2000. In a carefully thought-out paper, Tanaka admitted that nothing harbingers a sizeable market in the future, unless research activities come up with special products which bear fruit overnight. The scientists at Tanaka see replacement of gold as the greatest threat on all fronts. The only comfort they could offer was that at least the field of integrated circuits, transistors, connectors and printed circuit boards would grow, so that gold use might just stand still (with more miniaturisation and some replacement). On relays and switches they were more pessimistic, anticipating both a waning market and replacement of gold.

One discovery, which might slow the retreat from gold, is a gold-tin 'film', developed by Degussa in West Germany, which is ideal for plating integrated circuit boards. The alloying with tin makes the film cheaper than conventional gold-plating solutions (but, of course, continues to dilute the actual amount of gold needed).

The drive to replace gold seems to take little account of its apparent cost-effectiveness thanks to durability, but stems from the original cost compared to other metals; gold at $400 an ounce looks more forbidding than palladium at $140, silver at $7, or nickel at $0.15. Moreover, the rapid pace of technological advance means that many electronic consumer goods have a short life. Why use everlasting gold in a home computer or pocket calculator that will be superceded by a new model in two or three years? As miniaturisation progresses, incidentally, the sole consolation on the gold supply front is that the wisp of gold required for integrated circuits and printed circuit boards is scarcely worth recovering as scrap when they are discarded. Previously, thicker bonding wires and plating were worth recovering from obsolete equipment.

Satisfactory substitutes for gold are not found overnight, but the long-term view must be that consumption in electronics will remain stagnant for the next few years, at perhaps 100 to 120 tonnes annually, but during the 1990s will decline further. The fall could come more swiftly. If the US economy goes into recession, or US or European trade barriers resist Japanese electronics goods, gold would be an immediate casualty.

The outlook for dental gold is no brighter. The challenge here is on two fronts: increasing resistance from social security to pay for gold alloys, and competition from new palladium and ceramic-based alloys. Looking back over the last few years, the only surge in dental gold demand has come when governments changed the rules to permit gold alloys to be provided under health insurance. It happened first in Sweden between 1973 and 1974, when gold use doubled in a single year. When West Germany followed suit a couple of years later, dental gold offtake shot up 150 per cent. In fact, by 1979 West Germany accounted for over one-third of all dental gold consumption world-wide. Fabrication of alloys has even shot up in such diverse places as Switzerland and South Africa because dental companies were exporting to Germany. The authorities, faced with a bill at the peak for nearly $500 million for gold content alone, soon tightened up on the amount they would reimburse, and by 1986 patients were able to claim back only for the treatment itself, and not the gold alloys. Consequently, dental consumption in West Germany has fallen from a peak of around 30 tonnes, to just over 10 tonnes, and is forecast to decline even further before 1990.

Without Germany dental gold is a somewhat sorry picture. Only Japan shows slight improvement. But even there the challenge from palladium-based alloys is stunting any significant growth.

The one bright hope on the dental front is a gold powder paste, developed by Degussa, the West German fabricators, which can be compacted into solid gold in a tooth cavity by an ultra-sonic probe. The gold powder itself had been available for some time, but previously could only be compacted in the tooth by extensive hammering, much to the discomfort of patients. The ultra-sonic probe removes that misery. Whether the new technique, which awaits official medical approval, can even halt the decline in dental gold is doubtful. Certainly the gold mines cannot count on dentists to use up all their new production.

Specialist industrial and decorative uses for plating anything from costume jewellery to watch cases or bath taps, liquid gold for ceramic or label design, gold thread for saris, or gold leaf for gilding the domes of American state legislatures, or adorning statues of Buddha, are not growing either. Such applications called for about 60 tonnes annually twenty years ago, and scarcely 55 tonnes today. Plating solutions for jewellery, cigarette lighters, pens and pencils, watch cases and bracelets are all applied more sparingly; instead of 25 microns thick, the gold plate is 5 microns. Promising new techniques, developed mainly by Degussa in West Germany, for the direct plating of stainless steel with a sheen of gold-cobalt and a gold-iridium 'bath' to give a subtle soft gold hue to plated cutlery, do not add up to much actual consumption. The story is similar with gold-coated windows to reflect the sun in new office buildings, thus cutting air-conditioning bills. The potential looks good. Royal Bank of Canada used 77.7 kilos of gold in glazing the windows of their new Toronto skyscraper, and reportedly saved $50,000 annually in energy bills. The Gold Institute in Washington DC, which is sponsored by major gold miners and fabricators to campaign for technical and industrial uses of gold, came up with a Gold in Architecture Award to encourage such designs. The first went to AT & T for its new communications headquarters at Oakton, Virginia, where a long galleria has an entire ceiling in gold-coated glass. Golden glass houses, however, call for kilos of gold, not tonnes.

Medical applications remain on the periphery of gold use. The best-known is in the treatment of rheumatoid arthritis. The standard treatment has been to inject a soluble gold salt solution into the muscles. New research, however, has developed Auranofin, a gold compound which can be taken orally daily, and may avoid some of the unpleasant side effects of injected gold.[3] In India, the Gold Control Administration also continues to allocate about 75 kilos a year for Ayurvedic medicine, in which gold preparations are prescribed for 'heart and nerve tonic', melancholia, hardening of the arteries and even as an aphrodisiac.

The prospect, therefore, is that electronic, industrial, dental, decorative and medical uses of gold, as diverse and sometimes entertaining as they may be, are not a growth area. Indeed, looking ahead, the combined applications cannot be counted on for much more than 200 tonnes annually for the rest of this century, and possibly less if gold replacement proceeds rapidly in electronics.

Coupled with the uncertain future for jewellery, the message is that while the gold miners are forging ahead, the gold fabricators are not. Very radical initiatives are necessary if the 'bread-and- butter' demand is to hold its own as the safety net for the gold price. Meanwhile, the investor is being asked to shoulder a great deal.

3. Blain M. Sutton, 'Gold Compounds for Rheumatoid Arthritis', *Gold Bulletin*, Vol.19, No.1, 1986

Conclusion

THE VIEW TO 2000

The prospect for gold depends very much on your viewpoint. This book has not been about the short term of minutes or hours that often pre-occupies the speculator or professional trader. Rather, it is a snapshot of the gold business in the late 1980s, and tries to sense its direction for the rest of the century. It is less about where the price may be at any given moment, and more about the long-range underlying forces.

Judging those forces is not easy. The gold market's mood is fickle. The price was under $350 an ounce when I began this research and over $450 when I finished a year later. Sentiment has undergone a sea change. The debate was not how to absorb all the new production, but how tiny the market was compared to the liquidity in world currency, stock and bond markets. A technical chartist friend confided that he was looking for $1,100 an ounce sometime in 1989 or 1990 at the peak of the bull market.

The central long-term issue, however, has not changed. Travelling the world in pursuit of gold, what really stuck in my memory was not so much the fresh anxieties about the international economy, which have made gold a haven for investors again, but the new era in gold mining. The sustained excitement in gold is not in the markets, but in the outback of Australia, the jungles of Brazil, the Arctic of Canada and the deserts of Nevada. Gold mining has not had such vitality, nor been so profitable, at any other time in this century. And the momentum is still there. The most instructive thing in visiting the new gold fields was not just to see what was being done today, but to realise what is still untapped for tomorrow. Standing on a hill top in Nevada along the 'Carlin trend', the unfolding potential is apparent. In that vast landscape, Newmont Gold has so far pinpointed eleven orebodies; but they are pin-pricks on the map. Scores, perhaps hundreds, more satellite orebodies await discovery just on the spread of Newmont's T-Lazy-S Ranch. While in Australia and around the Pacific Basin's 'rim of fire', mining companies are only starting to get to grips with some huge epithermal deposits that will keep them busy well into the next century. Reckoning precisely how much gold will result is impossible, but one can say with certainty that the market will have to absorb much more gold each year for the rest of this century than could possibly have been forecast even five years ago.

The stimulant to mining has grown even stronger as the bull market that started in March 1985 gained strength. Gold mining was already profitable at $300-350 an ounce; with the price comfortably over $400 for the first half of 1987, it was even more attractive. As George Milling-Stanley observed in Consolidated Gold Fields' report, *Gold 1987*, 'Cost data for 1986 suggests that most operations increased their profit margins even further ... due to a combination of a higher gold price and a lowering of production costs at the mine sites.' Gold Fields calculates that in 1985 (the last year for which full data is available) net cash operating costs in the non-Communist world were on average $169 and total costs (including depreciation, finance charges and royalties) were $214.[1] An average price in the $420-450 range for 1987 represents a gravy train.

All good things come to an end. Mining companies are the last to complain about a higher gold price but, as an executive conceded, 'we are just storing up trouble for the future'. The problem is still how, or rather at what price, gold will be absorbed in the long term. Mining companies are under no illusions; they still look for prospects that can be brought into production with operating costs under $200 an ounce. They are finding them.

On the other side of the gold equation, however, the story is different. The underlying physical demand for gold in jewellery, industry and even coin has stagnated in the 1980s. Jewellery is still the largest single slice of the demand cake, but a much slimmer one than a decade ago. A consistent gap is opening up between the new supplies coming on to the market and the regular demand for fabrication, plus regional bar hoarding. The table opposite indicates how the underlying balance has changed, and the increasing emphasis that will have to be put on regular portfolio investment offtake (or central bank purchases) in the years ahead. The forecasts on the demand side will obviously depend on price fluctuations. The higher figures for jewellery or regional bar hoarding would require a price around $400 in 1987 dollars. The lower figures anticipate a price close to $500; on a sustained price over $500, the physical offtake would be substantially less than is forecast here, and the burden shifted to portfolio investment accordingly greater. Much more scrap would also come back over $500.

The potential for variation is shown in the chart on the following page, which reflects the monthly pattern of world-wide physical demand since 1983. Although the fluctuations here are not immediately as striking as those in the chart on Dubai, Singapore and Hong Kong offtake in Chapter 13, they show how the global demand varies considerably from month to month depending on the price. Only a 'hard core' of demand of about 50-70 tonnes monthly is not price-sensitive; set against new supplies averaging 130-40 tonnes a month, this demonstrates how the balance of the market can swiftly change. One month phys-

1. George Milling-Stanley, *Gold 1987*, Consolidated Gold Fields plc, London, May 1987

Gold Supply/Demand
(Excluding recycled scrap)

	1976 tonnes	1986 tonnes	1987 (E) tonnes	1988 (E) tonnes	1990 (F) tonnes	1995 (F) tonnes	2000 (F) tonnes
A. Supply							
Western mine	964	1281	1380	1450	1530	1450-1550	1450-1550
Communist sales	412	402	250-350	250-350	300-400	300-400	300-400
Total New Supply	1376	1683	1630-1730	1700-1800	1830-1930	1750-1950	1750-1950
B. Physical Demand							
Jewellery	936	828	750-800	750-800	750-850	750-900	750-900
Industry	217	230	220	220	210	210	200
Coin/medal	233	328	150-200	150-200	150-200	150-250	150-250
Regional bar investment/ hoarding	185	220	110-200	100-200	100-200	100-250	100-250
Total	1571	1606	1220-1420	1220-1470	1210-1460	1210-1610	1200-1650
A-B balance for portfolio investment/central banks	-195	77	210-510	230-580	370-720	140-740	100-750

E = estimate
F = forecast

ical offtake can far exceed conventional supplies (as in early 1985 on a low dollar price, or in mid-1986 on peak Japanese demand), and the next month fall far short. In fact, if Japan coin deliveries were excluded, the net hard-core offtake in October 1986 was a mere 14 tonnes, because so much dishoarding was coming from the Middle East and South-East Asia.

Physical Gold Demand – January 1983 to May 1987
GLOBAL

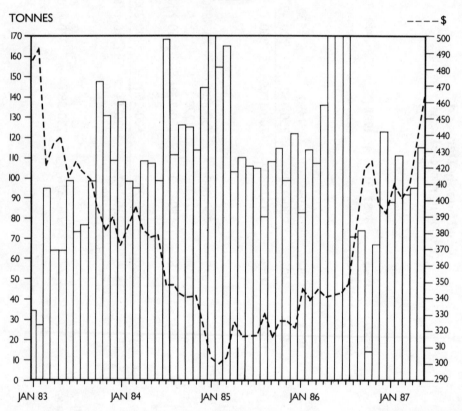

The scenario implied by the initial table is that the investor or some other special buyer is going to have to take a slice of the cake all the time. Japan saved the day in 1986, just as new supply and a higher level of communist sales were beginning to bite. The return of the investor the following year provided the next injection of funds, as the economic cycle swung back in favour of gold.

The humorous columnist Art Buchwald wrote a column during the 1980 bull market in which hé observed that while Americans wish each other, 'Have a good day,' gold traders wish each other, 'Have a bad day.' When the spectre of inflation returns, when Citicorp has to write off $3 billion in loan provisions for

Third World debts and Paul Volcker retires from the Federal Reserve, gold is again the beneficiary. A market price made by headaches may be good if you are selling aspirin, but gold cannot be forever dependent on them. The hang-over, to change the metaphor slightly, from the high price of 1980 lasted a long time; jewellery fabrication has never recovered to its original level. It should not be forgotten, either, that the real demand for silver in industry was seriously damaged by the $50 price, because the incentive to look for an alternative became imperative. A high price causes a high toll.

Gold is equally vulnerable. A major price rise would stimulate even more concentrated research to replace gold in electronics. Once the alternative is found, there is no going back. It would also erode the jewellery sector further. In fact, the 'safety net' of physical demand could be lowered, not raised. That threat cannot be avoided.

The speculator, eager to push the price onwards to $1,000 or beyond, is an integral part of today's market. The leverage available to him in futures and options trading can dictate the price and push it to limits never conceivable in the plebeian world of physical gold. In such heady moments, a few hard truths must be kept in mind. Gold may out-perform everything else for an afternoon, for a few weeks or even months. But it cannot be expected to do so indefinitely.

The great strength of gold in history has been to maintain its purchasing power over long periods. People have their favourite bench-marks: the price of a suit or a meal. Julian Baring at James Capel in London charts his 'gourmet's guide' to the gold price by the number of people who can dine at the Savoy Hotel in London for a sovereign. He starts at 1914, which was the last occasion on which a paper one pound note could be exchanged at the Bank of England for one sovereign. The set dinner cost £0.30 in 1914, so a sovereign bought 3.3 dinners; by 1987, the Savoy dinner cost £22 and the sovereign at £66 (with gold at $454) bought precisely three dinners. The relationship remained close and implied that gold was slightly under-valued, in sterling terms anyway. Taking a more broadly based view of gold's long-term purchasing power from 1934, when the $35 price level was decreed, to 1987, the price should be a shade under $300 per ounce if it had merely held its value in real terms. Thus, trading in a range of $400-500 gold was, in fact, turning in a better performance.

The interesting point is that when the price was around $300 a couple of years earlier, the safety net of physical demand was firmly in place. The grass roots buyer of bars and investment jewellery in the Middle East and South-East Asia felt intuitively that gold was cheap.

This is one part of the message. It signals that gold has re-established a realistic floor at close to its historical level of purchasing power. At $35, in 1970, it was hopelessly under-valued; at $850, a decade later, it was over-valued and people had lost all track of where the price ought to be. It took most of the 1980s to bring about a sense that $300-400 was a realistic price for miners and fabricators to work with, and for investors to have some confidence to get involved again. The word 'insurance' actually started to come back into conversations with portfolio managers in Switzerland.

But can the price build on that floor in real terms for the rest of this century? Here, I believe, the message is less auspicious. For the gold price to improve in line with inflation is one thing, but to pull ahead in real terms is quite another. That is to say, will it buy you more meals or a better-quality suit in the year 2000 than today? Undoubtedly, on market peaks, it will. But to buy an American Eagle and automatically expect it to buy you more services or goods in the year 2000 is another matter. The 'catching up' that the gold price had to do in the 1970s has been accomplished. Any sustained advance in real terms now would have to come from a genuine expansion of physical demand and permanent support from investors.

What does that require? Thinking ahead to the year 2000, it is worth remembering how jewellery has slipped. If we take 1986 as a pivotal year, and reflect back fourteen years and then on fourteen years, the challenge is clear. In 1972, jewellery consumed 995 tonnes or 85 per cent of Western mine production; in 1986, it was only sixty-four per cent. Projecting ahead to the year 2000 and assuming Western production at 1,450 tonnes, then jewellery has to be up to 928 tonnes just to stay level with 1986, or 1,218 tonnes (higher than it has ever been) to get back to the 1972 ratio.

Such an improvement can only be achieved, in my view, through bold and imaginative decisions by the mining industry on marketing and promotion. Nor is this just a question of advertising budgets, although money channelled thoughtfully into jewellery promotion can help. The real thrust has to be firmer strategies on high-carat, low mark-up jewellery, the structure of the jewellery industry and presentation to the customers. Jewellery has got to compete more effectively for the consumer dollar against clothes, electronics goods, home improvements and vacations. My confidence that something can be done is always bolstered in the Middle East or South-East Asia when I see visiting Americans, Europeans or Japanese descending on the jewellery stores and buying, for the first time in their lives, an ounce of 22-carat or *chuk kam* jewellery at close to the day's gold price.

The task, it must be admitted, is not an easy one. It is not only a matter of tapping a better share of consumer spending. Crime is a deterrent in many cities, especially in Latin America. It simply is not safe to go out wearing gold chains or earrings; they may be literally torn off. Women in Caracas or Rio de Janeiro go to parties with their jewellery in a grocery bag. That hardly helps sales.

The jewellery industry will have to run hard just to stand still. The torch, therefore, is passed to the elusive investors. If the gold price is going to be maintained, let alone show real advance, they ought to become a more permanent feature of the marketplace. The trouble is they tend to be increasingly transitory, because the nature of the game has changed. Gold is less and less bought to be squirrelled away, and more and more is traded to make money. The task, for gold's promotors at least, is to tap more of the huge pool of investment funds on a regular basis. This might lay the foundations of the recipe for a bullish scenario. For all its reputation, gold takes up a tiny part of people's savings, especially in Europe, North America and Japan. Shearson-Lehman, the investment bankers, have calculated that the total volume of gold investment from 1975 through 1985 was only 0.45 per cent of the capital invested in world stock markets, or 0.3 per cent of bond markets. That has to be the money to go for. Gold has got to compete with all those other seductive financial instruments against which it sometimes seems boring in comparison.

Trading papers may be fun, but it can also be dangerous. Governments can, and do, suspend markets overnight. What can you do with a futures contract if the exchange is closed? This is not to tell doomsday stories about taking to the hills with a sub-machine gun, a bicycle and a Krugerrand, but to point out that paper markets (and paper money) can be transitory. While writing this chapter, a gold dealer in Singapore telephoned to say that despite a high gold price ($462) there was a sudden rush for kilo bars in Indonesia. Why? Rumours, he said, of further devaluation of the rupiah. And it reminded me of a remark by Dr Henry Jarecki, the psychiatrist-turned-bullion-trader at Mocatta Metals in New York, who said, 'There is a strong sense in the stomachs of the people of the world that gold is good to hold.'

That sums it up. What gold still has going for it, even in the era of the electronic marketplace, is that it remains the only universally accepted medium of exchange. It is the lifebelt for all seasons, especially the dangerous ones. That credential has not been dented by futures or options or options on futures. The gold price, too, will have its seasons. The climate looks good for the late 1980s, but the clouds may gather in the early 1990s. A trough could then settle on the price for quite a while. But investors should not lose their nerve. The day of a volatile gold price is comparatively new. Like floating exchange rates, the floating gold price is not yet twenty years old. People often expect the price to perform miracles every week, when previously the miracle was that it did not change for as much as two centuries at time. Think of gold still as the golden 'constant'. And watch the mood of the *souks*; when they buy, gold is 'cheap'; when they sell, it is in speculative territory. What other signpost is needed?

Appendix I

Gold Supply to the Non-Communist Private Sector (tonnes)

	Western World Mine Production	Net Communist Sales (Purchases)	Net Official Sales (Purchases)	Old Gold Scrap	Total
1950	755	N/A	(288)	N/A	467
1951	733	N/A	(235)	N/A	498
1952	755	N/A	(205)	N/A	550
1953	755	67	(404)	N/A	418
1954	795	67	(595)	N/A	267
1955	835	67	(591)	N/A	311
1956	871	133	(435)	N/A	569
1957	906	231	(614)	N/A	523
1958	933	196	(605)	N/A	524
1959	1000	266	(671)	N/A	595
1960	1049	177	(262)	N/A	964
1961	1080	266	(538)	N/A	808
1962	1155	178	(329)	N/A	1004
1963	1204	489	(729)	N/A	964
1964	1249	400	(631)	N/A	1018
1965	1280	355	(196)	N/A	1439
1966	1285	(67)	40	N/A	1258
1967	1250	(5)	1404	N/A	2649
1968	1245	(29)	620	N/A	1836
1969	1252	(15)	(90)	N/A	1147
1970	1273	(3)	(236)	N/A	1034
1971	1233	54	96	N/A	1383
1972	1177	213	(151)	N/A	1239
1973	1111	275	6	N/A	1392
1974	996	220	20	N/A	1236
1975	946	149	9	N/A	1104
1976	964	412	58	N/A	1434
1977	962	401	269	N/A	1632
1978	972	410	362	N/A	1744
1979	959	199	544	N/A	1702
1980	959	90	(230)	482	1301
1981	981	280	(276)	232	1217
1982	1028	203	(85)	237	1383
1983	1115	93	142	289	1639
1984	1160	205	85	284	1734
1985	1233	210	(135)	299	1607
1986	**1281**	**402**	**(181)**	**465**	**1967**

Definition of official sales was extended from 1974 to include activities of government-controlled investment and monetary agencies in addition to central bank operations. This category also includes IMF disposals.
Old gold scrap figures are only available from 1980. 'N/A' means 'Not available'.
Source: *Gold 1987*, Consolidated Gold Fields plc.

Appendix II

Mine Production in the non-Communist World (tonnes)

	1980	1986	1987(E)	1990(E)	1995(E)	2000(E)
Europe	11.8	16.5	16.5	16.5	16.5	16.5
North America						
United States	30.5	108.5	140.0	175.0	150.0	150.0
Canada	51.6	107.5	115.0	150.0	120.0	120.0
Total North America	82.1	215.5	255.0	325.0	270.0	270.0
Latin America						
Brazil	35.0	67.4	65.0	70.0	75.0	85.0
Colombia	17.0	27.1	25.0	25.0	20.0	20.0
Chile	6.5	19.2	21.0	19.0	15.0	15.0
Venezuela	1.0	15.0	25.0	30.0	25.0	20.0
Peru	5.0	10.9	10.0	12.0	10.0	10.0
Dominican Republic	11.5	9.1	8.0	7.0	6.0	5.0
Mexico	5.9	8.3	8.0	9.0	10.0	15.0
Bolivia	2.0	6.0	6.0	7.0	8.0	10.0
Other	4.8	7.5	8.0	14.0	15.0	20.0
Total Latin America	88.7	170.5	176.0	193.0	184.0	200.0
Saudi Arabia	—	—	—	3.5	34.5	3.5
India	2.6	2.1	2.1	2.0	3.0	3.0
Far East						
Philippines	22.0	39.9	37.0	33.0	30.0	30.0
Japan	6.7	14.0	14.0	15.0	15.0	15.0
Other (incl. Indonesia)	4.5	14.6	20.0	25.0	40.0	50.0
Total Far East	33.2	68.5	71.0	73.0	85.0	95.0
Africa						
South Africa	675.1	640.0	650.0	680.0	690.0	680.0
Zimbabwe	11.4	14.9	15.0	15.0	15.0	20.0
Ghana	10.8	11.5	12.0	18.0	18.0	20.0
Zaire	3.0	8.0	8.0	8.0	10.0	10.0
Other	8.0	18.2	17.0	20.0	25.0	25.0
Total Africa	708.3	692.6	702.0	741.0	758.0	755.0
Australasia						
Australia	17.0	75.0	115.0	145.0	110.0	100.0
Papua New Guinea	14.3	36.1	45.0	47.0	60.0	60.0
Other (incl. N. Zealand)	1.0	4.0	4.0	7.0	12.0	20.0
Total Australasia	32.3	115.1	164.0	199.0	182.0	180.0
TOTAL	959.0	1280.8	1386.0	1553.0	1502.0	1523.0

Source: *Gold 1987* and author's estimates

Appendix III

South Africa – Top 20 Gold Producers by the 1990s

	Mine	Major Share Holders	Output in tonnes 1987	1990
1	Freegold	Anglo American	108.5	105.0
2	Vaal Reefs	Anglo American	80.6	84.0
3	Driefontein Cons	Gold Fields of SA	60.0	60.0
4	Western Deeps	Anglo American	40.0	60.0
5	Kloof	Gold Fields of SA	31.6	37.0
6	Harmony	Barlow Rand	32.9	34.0
7	Buffelsfontein	Gencor	34.8	32.5
8	Leslie	Gencor	3.4	32.0
9	Randfontein	Johannesburg Cons Investments	27.2	31.0
10	Hartebeesfontein	Anglovaal	31.2	30.0
11	West Rand Cons	Gencor	3.9	30.0
12	Western Areas	Johannesburg Cons Investments	16.0	22.0
13	Elandsrand	Anglo American	15.0	20.0
14	Winkelhaak	Gencor	13.7	13.0
15	Deelkraal	Gold Fields of SA	7.6	12.0
16	Kinross	Gencor	12.8	12.0
17	Blyvooruitzicht	Barlow Rand	12.0	11.0
18	Doornfontein	Gold Fields of SA	8.6	9.0
19	Unisel	Gencor & Seltrust (BP Minerals)	9.5	9.0
20	Loraine	Anglovaal	8.8	9.0

USA – Top 20 Gold Producers by the 1990s

	Mine	Major Share Holders	Output in tonnes 1987	1990
1	Gold Quarry Complex	Newmont	12.6	16.0
2	Homestake	Homestake	11.9	11.8
3	Round Mountain	Echo Bay, Homestake, Case Pomeroy	5.3	10.0
4	Fortitude (Battle Mt)	Battle Mountain Gold	7.2	9.5
5	Jerritt Canyon	Freeport-McMoRan, FMC Gold	8.5	8.0
6	Carlin	Newmont	5.0	5.0
7	Chimney Creek	Cons Gold Fields	0.0	5.0
8	McLaughlin	Homestake Mining	5.5	5.0
9	Mesquite	Cons Gold Fields	4.6	4.5
10	Jamestown	Sonora Gold	3.8	4.0
11	Pine Tree (Josephine)	Goldenbell Resources	0.0	4.0
12	Cannon	Asamera, Breakwater Resources	4.0	3.7
13	Montana Tunnels	Pegasus Gold	1.9	3.6
14	Mercur	American Barrick Resources	3.7	3.1
15	Rain	Newmont	0.0	3.0
16	Golden Sunlight	Placer	2.9	3.0
17	Summitville	Galatic Resources	2.5	2.7
18	McCoy	Echo Bay Mines	2.5	2.6
19	Rawhide	Kiewit, Kennecott, Plexus Res	0.6	2.4
20	Zortman/Landusky	Pegasus Gold	2.5	2.3

Canada – Top 20 Gold Producers by the 1990s

	Mine	Major Share Holders	Output in tonnes 1987	1990
1	Hemlo Page-Williams	Lac Minerals or International Corona?	8.0	15.0
2	Hemlo Golden Giant	Noranda, Golden Sceptre, Goliath	10.0	11.0
3	Doyon	Lac Minerals, Cambior	7.2	7.8
4	Campbell	Dome Mines	7.1	7.5
5	Pamour Porcupine Mill & Tails	Jimberlane Holdings	3.8	6.5
6	Lupin	Echo Bay Mines	6.2	6.2
7	Bousquet (& Bousquet East)	Lac Minerals	2.5	6.0
8	Hemlo David Bell	Teck Corp, Inter. Corona Resources	3.7	5.0
9	Cinola	City Resources, Energy Resources	0.0	4.5
10	Golden Knight (Casa Barardi)	Inco, Golden Knight Res. Teck	0.0	4.5
11	Belmoral	Belmoral Mines	2.4	4.0
12	Hope Brook (Chetwynd)	BP Minerals	0.8	4.0
13	Dome	Dome Mines	4.0	4.0
14	Macassa (Willroy)	Lac Minerals	2.2	3.5
15	Detour Lake	Dome Mines, Amoco	1.9	3.5
16	Holt-McDermott	American Barrick Resources	0.0	3.0
17	Con	Nerco	3.0	3.0
18	White (Red Lake)	Dickenson Group, Sullivan Mining	2.6	3.0
19	Nickel Plate	Royex Gold Mining	1.0	3.0
20	Copper Rand/Portage	Northgate Exploration	2.4	2.5

Australia – Top 20 Gold Producers by the 1990s

Mine	Major Share Holders	Output in tonnes 1987	1990
1 Big Bell	ACM, Placer Pacific	0.0	7.5
2 Mt Charlotte/Fimiston	KMA (Western Mining), Homestake	7.2	7.0
3 Kidston	Placer Pacific	7.0	7.0
4 Kambalda	Western Mining	3-4	6.0
5 Boddington	Reynolds, Shell, BHP, Kobe	0.0	5.0
6 Fimiston/Paringa	North Kalgurli	4.0	4.0
7 Telfer	Newmont, BHP	5.2	4.0
8 Lady Bountiful	WMC, Consolidated Exploration	0.8	3.5
9 Norseman	Australis	3.4	3.0
10 Paddington	Pancontinental	2.8	2.8
11 Mt Pleasant/Golden Kilometre	Elders, Square Gold	1.1	2.5
12 Tennant Creek	Peko Wallsend	2.5	2.5
13 Pajingo	Battle Mountain Gold	0.0	2.0
14 Red Dome	Elders Resources	1.1	2.0
15 Pine Creek	Renison Goldfields	2.0	2.0
16 Mt Morgan	Peko Wallsend, Anglo American	1.7	2.0
17 Harbour Lights	Carr Boyd, Aztec	1.8	1.8
18 Mt Wilkenson	Chevron	0.7	1.7
19 Emu	Western Mining	1.5	1.5
20 Mt Percy	Windsor, North Kalgurli	1.4	1.6

Pacific Rim – Top 20 Gold Producers by the 1990s

Mine	Major Share Holders	Output in tonnes 1987	1990
Papua New Guinea			
Porgera	Placer Pacific, Renison, MIM	0.0	24.0
Bougainville	CRA PNG Govt	13.3	17.0
OK Tedi	BHP, Amoco	35.0	13.0
Lihir	Kennecott Niugini	0.0	12.0
Misimi	Placer Pacific	0.0	5.0
Philippines			
Philex	Philex	5.6	7.0
Benguet Gold	Benguet Corp	3.9	4.0
Dizon	Benguet Corp	3.5	3.4
Atlas Masbate	Atlas Cons Mining & Dev Co	2.4	2.4
Atlas Cebu	Atlas Cons Mining & Dev Co	2.3	2.3
Lepanto	Lepanto	2.0	2.0
Philippine Eagle	Paragon Res	0.7	1.8
New Zealand			
Martha Hill/Waihi	Amax, ACM	0.1	2.0
Fiji			
Emperor	Western Mining Emperor Mines	2.6	3.0
Japan			
Hishikari	Sumitomo Metal Mining	6-7	6-7
Indonesia			
Masuperia	BP Minerals, Aneka Tambanq	0	5-10
Kelian	CRA	0.0	6.0
Kasongan	Pelsart, Jason	3	3-4
Ertsberg	Freeport, McMoran	3.0	2.7
Malaysia			
Mamut	OMR Dev	2.2	2.2

Latin America – Top 20 Gold Producers by the 1990s

Mine	Major Share Holders	Output in tonnes 1987	1990
Brazil			
Garimpos	Garimpeiros	57.0	52.0
Morro Velho	Anglo American Bozamo Simonsen	5.0	5.0
Crixas	Anglo American, INCO	0.0	4.0
Morro Do Ouro	RTZ, Antram	0.4	3.4
Serra Pelada	Garimpeiros	2.0	2.0
Jacobina	Anglo American	1.2	2.4
Sao Bento	Gencor, Amira Trading	1.5	2.1
Cabacal/Santa Martha	BP, COBEM	0.8	2.0
Novo Astro	GMP Group	0.5	1.5
Transamazonica	Northern Territory Resources	0.2	0.9
Novo Planeta	Paranapanema	0.9	0.7
Mexico			
Tayoltita/La Libertad	Corp Indust Sanluis	1.5	3.0
Torres Complex	Cia Fresnillo, Lacana, Penoles	1.6	1.6
Paradones Amarillos	Imperial Metals Corp	0.0	0.5
Dominican Republic			
Pueblo Viejo	Rosario Dominicana	9.5	8.9
Chile			
El Indio	St Joe	9.0	6.0
Hueso de Silica	Codelco	1.0	2.0
Guyana			
Omai	Placer, Golden Star Resources	0.0	2.3
Colombia			
Antoquia	Mineros de Antoquia	1.3	1.3
Venezuela			
Mineros	Mineros	20-25	30-35

Arabia to Zimbabwe – Top 20 Gold Producers by the 1990s

Mine	Major Share Holders	Output in tonnes 1987	1990
Ghana			
Ashanti	Ghana Gov, Lonhro	9.7	12.0
Tarkwa	State Gold Mining Corp	0.6	1.3
Obenemase	State Gold Mining Corp	0.0	1.2
Prestea	State Gold Mining Corp	0.5	0.6
Zaire			
Kilo-Moto	Gold Mines of Kilo-Moto	3.0	6.0
Zimbabwe			
Dalny/Venice	Falcon Mines	2.1	1.5
Redwing/Old West	Lonhro	0.8	0.8
Vubachikwe	Forbes/Thompson (PVT)	0.8	0.8
How	Lonhro	0.8	0.8
Rebecca/Freda	Cluff Oil Holdings	0.0	0.7
Sudan			
Gebeit	Greenwich Resources, Govt	1.3	2.5
Saudi Arabia			
Mahd Ad-Dhahab	Petromin, Cons Gold Fields	0.0	2.0
Sukhairat	Boliden Govt	0.0	1.5
Ivory Coast			
Asupiti/Aniuri	Eden Roc Mineral Group	0.0	2.0
African Alluvial	Various	25.0	20-25

Appendix IV

The Major Gold Mining Groups by the 1990s

Company	Mine	Location	% share
Anglo American Corp of South Africa	Elandsrand	South Africa	100
	Ergo	South Africa	100
	Freegold	South Africa	100
	Vaal Reefs	South Africa	100
	Western Deep Levels	South Africa	100
	Morro Velho	Brazil	70
	Jacobina	Brazil	70
	Crixas	Brazil	50
	Mt Morgan	Australia	40
Anglo Vaal	Hartebeesfontein	South Africa	100
	Loraine	South Africa	100
	Village Main	South Africa	100
	East Tvl Cons	South Africa	100
Gencor Group	Barberton	South Africa	100
	Bracken	South Africa	100
	Buffelsfontein	South Africa	100
	Grootvlei	South Africa	100
	Kinross	South Africa	100
	Leslie	South Africa	100
	Marievale	South Africa	100
	St Helena	South Africa	100
	Stilfontein	South Africa	100
	Unisel	South Africa	66
	West Rand Cons	South Africa	100
	Winkelhaak	South Africa	100
	Sao Bento	Brazil	49
Barlow Rand Group	Blyvooruitzicht	South Africa	100
	Durban Deep	South Africa	100
	ERPM	South Africa	100
	Harmony	South Africa	100
	Rand Mines Props	South Africa	100

Johannesburg Consolidated Investments	Joel	South Africa	100
	Randfontein	South Africa	100
	Western Areas	South Africa	100
Consolidated Gold Fields plc (Owns 48% GFSA, 26% Newmont, 49% Renison)	Ortiz	USA	100
	Mesquite	USA	100
	Chimney Creek	USA	100
Gold Fields of South Africa (GFSA)	Deelkraal	South Africa	100
	Doornfontein	South Africa	100
	Kloof	South Africa	100
	Libanon	South Africa	100
	Vlakfontein	South Africa	100
	Venterspost	South Africa	100
	Driefontein Cons	South Africa	100
Newmont Mining Corporation	Carlin	USA	90
	Maggie Creek	USA	90
	Blue Star	USA	90
	Gold Quarry	USA	90
	Rain	USA	90
	Telfer	Australia	70
	New Celebration	Australia	50
Renison Goldfields Consolidated	Pine Creek	Australia	60
	Porgera	Papua New Guinea	33
	Wau	Papua New Guinea	100
Lac Minerals Ltd (awaiting legal decision on Page Williams – see Inter Corona)	Page Williams	Canada	?
	Doyon	Canada	50
	Bousquet	Canada	100
	Lake Shore Macassa	Canada	100
Barrack Mines Ltd	Horseshoe Lights	Australia	66
	Wiluna	Australia	25
	Croydon	Australia	92
American Barrick Resources Corp	Mercur	USA	100
	Goldstrike	USA	100
	Camflo	Canada	100
	Pinson	USA	26
	Renabie	Canada	50
	Valdez Creek	USA	23
	Holt-McDermott	Canada	100

Royex Gold Mining Corp (owns 49% of Corona)	Mascot (Nickel plate)	Canada	52
	Renabie	Canada	50
	Mallard Lake	Canada	50
International Corona Resources Ltd (awaiting legal decision on Page-Williams – see Lac Minerals)	David Bell	Canada	50
	Page Williams	Canada	?
Placer Dome Inc (Placer, Dome and Campbell Red Lake merger announced)	Kidston	Australia	70
	Golden Sunlight	USA	100
	Cortez	USA	40
Placer Development/ Placer Pacific	Bald Mountain	USA	84
	Big Bell	Australia	50
	Misima	Papua New Guinea	100
	Porgera	Papua New Guinea	31.3
Dome Mines	Campbell Red Lake	Canada	100
	Detour Lake	Canada	50
	Dona Lake	Canada	100
Pegasus Gold Inc	Zortman/Landusky	USA	100
	Florida Canyon	USA	100
	Montana Tunnels	USA	100
	Relief Canyon	USA	100
Western Mining Corp Holdings Ltd	Emperor	Fiji	20
	Lady Bountiful	Australia	50
	Emu	Australia	100
	Kambalda	Australia	100
	Lancefield	Australia	100
	Central Norseman	Australia	51
	Stawell	Australia	72.5
	Hill 50	Australia	38
Broken Hill Pty Co (BHP)	Boddington	Australia	20
	Browns Creek	Australia	100
	Ora Banda	Australia	100
	OK Tedi	Papua New Guinea	31.2
	Telfer	Australia	30
	Mt Charlotte/Fimiston	Australia	52

Freeport-McMoran Inc	Jerritt Canyon	USA	70
	Getchell	USA	100
	Ertsberg	Indonesia	85
Australian Consolidated Minerals Ltd (ACM)	Big Bell	Australia	50
	Golden Crown	Australia	100
	Westonia	Australia	100
	Waihi	New Zealand	n/a
Echo Bay Mines Ltd	Round Mountain	USA	50
	Sunnyside	USA	100
	Lupin	Canada	100
	McCoy	USA	100
	Manhattan	USA	100
	Borealis	USA	100
Homestake Mining Co	McLaughlin	USA	100
	Homestake	USA	100
	Round Mountain	USA	25
	Mt Charlotte	Australia	48
	Fimiston	Australia	48
Battle Mountain Gold	Pajingo	Australia	100
	Fortitude	USA	100
Teck Corp	Hemlo David Bell	Canada	50
	Golden Knight	Canada	12

Appendix V

Gold Fabrication in Developed and Developing Countries (tonnes) (excluding the use of scrap)

	1980	1981	1982	1983	1984	1985	1986
Developed countries							
Jewellery	240.6	361.0	411.8	375.5	432.9	493.0	511.2
Electronics	93.5	91.5	87.3	104.8	128.2	111.6	119.4
Dentistry	62.5	63.7	59.1	50.0	51.0	51.5	49.3
Other industrial	57.5	57.8	54.0	49.5	52.2	49.0	51.3
Medals and imitation coins	18.1	12.4	5.8	22.3	15.7	3.4	4.2
Official coins	169.7	141.5	123.6	152.2	124.3	91.0	300.9
Sub-total	641.9	727.9	741.6	754.3	804.3	799.6	1,036.3
Developing countries							
Jewellery	40.2	251.0	320.0	254.2	422.0	398.9	316.5
Electronics	1.8	1.4	1.6	1.7	2.3	2.9	4.2
Dentistry	1.8	1.5	1.5	0.8	1.1	1.5	1.5
Other industrial	4.2	4.3	3.9	3.2	3.3	5.2	5.3
Medals and imitation coins	2.6	14.9	15.9	9.3	28.3	10.9	7.5
Official coins	20.5	49.8	7.0	12.9	6.2	13.5	25.7
Sub-total	71.1	322.9	349.9	282.1	463.2	432.9	360.2
TOTAL	713.0	1050.8	1091.5	1036.4	1267.5	1232.5	1396.5

Source: adapted from *Gold 1987* Consolidated Gold Fields
Developed countries: Western Europe, North America, Japan, South Africa, Australia

Appendix VI

Official Gold Holdings (tonnes)

	1970	1975	1980	1986
USA	9,839	8,544	8,221	8,150
Canada	703	683	653	622
Austria	634	649	657	658
Belgium	1,307	1,312	1,063	1,063
France	3,139	3,139	2,546	2,546
Germany	3,536	3,658	2,960	2,960
Italy	2,565	2,565	2,074	2,074
Japan	473	657	754	754
Netherlands	1,588	1,690	1,367	1,367
Portugal	802	862	690	628
South Africa	592	552	378	149
Switzerland	2,427	2,588	2,590	2,590
UK	1,199	654	583	592
Opec	1,042	1,084	1,230	1,362
Other Asia	545	487	612	1,075
Other Europe	1,024	1,126	1,212	1,311
Other Middle East	427	451	463	459
Other Western Hemisphere	646	578	663	678
Rest of the World	298	280	321	329
Unspecified	85	86	106	171
Total All Countries	32,871	31,645	29,541	29,526
IMF	3,856	4,772	3,217	3,217
EMCF	—	—	2,663	2,692

This gold accounts for drop shown in holdings of EEC countries above

Source: IMF

Appendix VII

Gold Price 1970-87

	Highest $	Lowest $	Year-End $	Average $
1970	39.19	34.35	37.38	35.96
1971	43.97	27.32	43.63	40.81
1972	70.00	43.72	64.90	58.20
1973	127.00	63.90	112.25	97.22
1974	197.50	114.75	186.50	158.80
1975	186.25	128.75	140.25	160.87
1976	140.35	103.05	134.51	124.79
1977	168.15	129.40	164.95	147.71
1978	243.65	165.70	226.00	193.51
1979	524.00	216.55	524.00	305.85
1980	850.00	474.00	589.50	614.63
1981	599.25	391.25	400.00	460.13
1982	488.50	296.75	448.00	375.64
1983	511.50	374.25	381.50	423.68
1984	406.85	303.25	309.00	360.72
1985	340.90	284.25	327.00	317.33
1986	442.75	326.00	390.90	368.03
1987 (Jan-May)	476.60	390.00	—	423.66

Bibliography

Beale, Robert; *Trading in Gold Futures*, Woodhead-Faulkner, Cambridge, Nichols Publishing Co., New York, 1985

Canadian Mining Handbook, Northern Miner, Toronto, 1986

Green, Timothy, *The New World of Gold*, Weidenfeld & Nicolson, London; Walker & Co., New York, 1981 (revised 1985)

Jastram, Roy, *The Golden Constant*, John Wiley & Sons, New York, 1977

Quadrio-Curzio, Alberto, *The Gold Problem: Economic Perspectives*, Oxford University Press for Banca Nazionale del Lavoro and Momisma, 1982

Register of Australian Mining 1987, Resource Information Unit, in association with Australian Business and the Perth Stock Exchange, Perth, Western Australia 1987

Sutherland, C.H.V., *Gold: its beauty, power and allure*, Thames and Hudson, London, 1959

The following publications have also been essential source material:

Annual Bullion Review, Samuel Montagu & Co., London

Aurum, World Gold Council, Geneva, Switzerland

Australian Business, Sydney, Australia

Chamber of Mines Newsletter, Johannesburg

Financial Times, World Gold Conferences, speakers' papers, 1985-7

Gold, Annual Surveys 1969-87, Consolidated Gold Fields plc, London

Gold Bulletin, International Gold Corporation, Johannesburg

Gold Gazette, Resource Information Unit, Perth, Western Australia

Jewellery News Asia, Hong Kong

Mining Survey, Chamber of Mines, Johannesburg, South Africa

Newsletter, Chamber of Mines of the Philippines

Northern Miner, Toronto

Precious Metals Market in Japan, 17th edition, March 1987, Sumitomo Corporation, Tokyo

Index